Multicriteria Optimization - Pareto-Optimality and Threshold-Optimality

Edited by Nodari Vakhania and Frank Werner

Published in London, United Kingdom

IntechOpen

Supporting open minds since 2005

Multicriteria Optimization - Pareto-Optimality and Threshold-Optimality
http://dx.doi.org/10.5772/intechopen.78897
Edited by Nodari Vakhania and Frank Werner

Contributors
Vladislav Zhukovskiy, Konstantin Kudryavtsev, Maznah Kasim, David Arditi, Ibraheem Alothaimeen, Nodari Vakhania, Alexander Lazarev, Nikolay Pravdivets

First published in London, United Kingdom, 2020 by IntechOpen
IntechOpen is the global imprint of INTECHOPEN LIMITED, registered in England and Wales, registration number: 11086078, 5 Princes Gate Court, London, SW7 2QJ, United Kingdom
Printed in Croatia

British Library Cataloguing-in-Publication Data
A catalogue record for this book is available from the British Library

Additional hard and PDF copies can be obtained from orders@intechopen.com

Multicriteria Optimization - Pareto-Optimality and Threshold-Optimality
Edited by Nodari Vakhania and Frank Werner
p. cm.
Print ISBN 978-1-78984-718-5
Online ISBN 978-1-78984-719-2
eBook (PDF) ISBN 978-1-83969-265-9

For EU product safety concerns:
IN TECH d.o.o., Prolaz Marije Krucifikse Kozulić 3, 51000 Rijeka, Croatia, info@intechopen.com or visit our website at intechopen.com.

Meet the editors

Nodari Vakhania graduated with honors from the Faculty of Applied Mathematics and Cybernetics, Tbilisi State University, Georgia, in 1983. In 1989, he entered the PhD program in Computer Science at the Russian Academy of Sciences, Moscow, and obtained a degree in Mathematical Cybernetics in 1991. In 1992, he was a postdoctoral fellow at the Russian Academy of Sciences, and had a short-term visiting position at the University of Saarbrucken, Germany. In 1995, he became a professor at the Centro de Investigación en Ciencias at the State University of Morelos, Mexico. He also has an honorary position at the Institute of Computational Mathematics of the Georgian Academy of Sciences, where he received a doctoral degree in Mathematical Cybernetics in 2004. His research interests include design and analysis of algorithms, discrete optimization, computational complexity and scheduling theory. He is an author of nearly 100 refereed research papers including more than 60 publications in highly ranked international journals. He has also worked with different scientific committees, including those at the Mexican Science Foundation CONACyT. He is an editorial board member and a referee for a number of international scientific journals. He has obtained research grants and honors in Germany, France, the Netherlands, the United States, Russia, and Mexico, and has given more than forty invited talks throughout the world.

Frank Werner studied Mathematics from 1975 to 1980 and graduated from the Technical University Magdeburg (Germany) with honors. He defended his PhD thesis in 1984 (summa cum laude) and his habilitation thesis in 1989. In 1992, he received a grant from the Alexander von Humboldt Foundation. Currently, he works as Extraordinary Professor at the Faculty of Mathematics of the Otto-von-Guericke University Magdeburg (Germany). He is an author or editor of seven books, and he has published more than 280 papers. He is on the editorial board of seventeen journals; in particular, he is editor-in-chief of *Algorithms* and an associate editor of *International Journal of Production Research* and *Journal of Scheduling*. His research interests include operations research, combinatorial optimization, and scheduling.

Contents

Preface

Multi-criteria optimization problems naturally arise in practice when there is no single criterion for measuring the quality of a feasible solution, a solution that satisfies the restrictions of the problem. Since different criteria are contradictory, it is difficult and often impossible to find a single feasible solution that is good for all the criteria, that is, the problem cannot be addressed as a common optimization problem (with a single objective criterion). Hence, some compromise is unavoidable for the solution of multi-criteria optimization problems. A commonly accepted such compromise is to look for a Pareto-optimal frontier of the feasible solutions (i.e., a set of feasible solutions that are not dominated by any other feasible solution with respect to any of the given criteria). Although this is a reasonable compromise, it has two drawbacks. Firstly, finding a Pareto-optimal frontier is often computationally intractable, and secondly, a practitioner may be interested in solutions with some priory given (acceptable) value for each of the objective functions, not in just a set of the non-dominated feasible solutions. Hence, other effective optimality measures for the multi-criteria optimization problems are possible. Theoretical explorations of possible generalizations, relaxations and variations of standard Pareto-optimality principles may lead to robust, and at the same time, flexible and practical measures for multi-criteria optimization (with a "fair balance" for all the given objective criteria). In this book, besides the traditional Pareto-optimality approach (Section 1), we suggest one new alternative approach for the generation of an admissible solution to a multi-criteria optimization problem (Section 2). The book also presents two overview chapters on the existing solution methods for two real-life, multi-criteria optimization problems (Section 3).

The first chapter in Section 1 addresses multi-criteria problems that arise in game theory when each player has their own goal that does not coincide with the goal of the other players. The strategy of each player is measured by its payoff function, which value depends not only on the decisions made by the player but also on the decisions of the remaining players. To optimize their goal, the player needs to take into account possible actions of the other players. A game is traditionally referred to as noncooperative if different players cannot coordinate their actions between each other. For a noncooperative game, it is commonly accepted that the Nash equilibrium gives a reasonable solution for all the players, the so-called Nash equilibrium strategy profile. As the authors observe, two different profiles from the set of a Nash equilibrium strategy profiles might not be "equally good," that is, there may exist two different Nash equilibrium strategy profiles such that the payoffs of each player in the first strategy profile are strictly greater than the corresponding payoffs in the second one. Therefore, it is natural to look for a Nash equilibrium strategy that is Pareto optimal with respect to the rest of the Nash equilibrium strategies. The authors, in light of their earlier relevant results, expand their line of research in this direction and suggest new solution methods.

The second chapter of Section 1 considers the Pareto-optimality setting for a bi-criteria machine-scheduling problem. Broadly speaking, the scheduling problems deal with a set of jobs or orders that are to be performed by a set of resources or machines. There is a basic, traditional resource restriction that a machine can handle at most one job at a time, and there may be additional restrictions on the ways of how the jobs can be scheduled on the machines, which define a set of feasible solutions. Moreover, we have one or more objective functions defined on the set of feasible solutions that are to be minimized or maximized, as in any other discrete optimization problem. This chapter considers a single-machine environment and aims to minimize two objective functions: the maximum job lateness and the maximum job completion time. Since the problem of finding the Pareto-optimal set of feasible solutions for this bi-criteria problem is NP-hard, the authors consider a special case when a polynomial-time solution is possible.

The second section consists of a chapter that introduces an alternative optimality measure for multi-criteria optimization problems. The new approach is motivated by the observation that, in practical circumstances, there may exist different tolerance/requirements to the quality of the desired solution for different objective criteria. In other words, for some objective criteria, solutions that are far away from an optimal one can be acceptable, whereas for some other criteria, near-optimal solutions are required. Hence, a uniform homogeneous Pareto-optimality approach may not be good from the point of view of the practical needs, and the computational point of view, since most Pareto-optimality problems are known to be intractable. Even if the Pareto-optimal set of feasible solutions is created, it may not be computationally efficient to choose an appropriate solution from the Pareto-optimal set of feasible solutions. The multi-threshold optimization setting suggested in this chapter takes into account different requirements for different objective criteria. Hence, a single feasible solution with an admissible value for each objective function can be generated with the computational cost, inferior to the cost of finding the corresponding Pareto-optimal feasible set.

The last section consists of two survey chapters. The first overviews the multi-criteria optimization problems that arise in construction projects. In such projects, there are different contradictory criteria and it is a challenging question to meet all of them. Hence, multi-criteria optimization methods are applied for the solution of such problems. The chapter surveys the solution methods including different meta-heuristic and genetic algorithms and integer programming methods. The last chapter overviews some work that relates a practical problem of measuring an expert's credibility in evaluation of the importance of different criteria with the multi-criteria optimization problems. In particular, the situation when different criteria may have different importance (which is represented by the corresponding weights) is considered. Common methods for determining the relative importance of each criteria and the corresponding feasible solutions are briefly described.

Nodari Vakhania
Centro de Investigación en Ciencias,
UAEM,
Mexico

Frank Werner
Otto-von-Guericke University,
Germany

Section 1

Traditional Approach: Pareto Optimality

Chapter 1

Pareto Optimality and Equilibria in Noncooperative Games

Vladislav Zhukovskiy and Konstantin Kudryavtsev

Abstract

This chapter considers the Nash equilibrium strategy profiles that are Pareto optimal with respect to the rest of the Nash equilibrium strategy profiles. The sufficient conditions for the existence of such pure strategy profiles are established. These conditions employ the Germeier convolutions of the payoff functions. For the noncooperative games with compact strategy sets and continuous payoff functions, the existence of the Pareto-optimal Nash equilibria (PoNE) in mixed strategies is proven.

Keywords: Pareto optimality, Nash equilibrium, Pareto-optimal Nash equilibrium, noncooperative game, Germeier convolution

1. Introduction

In 1949, J. Nash, a Princeton University graduate at that time and a famous American mathematician and economist as we know him today, suggested the notion of an equilibrium solution for a noncooperative game [1] lately called "the Nash equilibrium strategy profile." Since then, this equilibrium is widely used in economics, sociology, military sciences, and other spheres of human activity. Moreover, 45 years later J. Nash, J. Harshanyi, and R. Selten were awarded the Nobel Prize "for the pioneering analysis of equilibria in the theory of noncooperative games."

However, as shown by Example 1, the set of the Nash equilibrium strategy profiles has a negative property: there may exist two Nash equilibrium strategy profiles such that the payoffs of each player in the first strategy profile are strictly greater than the corresponding payoffs in the second one. In 2013, the authors emphasized this fact in a series of papers [2, 3] while exploring the existence of a guaranteed equilibrium solution for a noncooperative game under uncertainty. Particularly, these papers were focused on the Nash equilibrium strategy profile that is Pareto optimal with respect to the rest of the Nash equilibrium strategy profiles, thereby eliminating the above shortcoming. And the following question arises immediately. How can such an equilibrium (the so-called Pareto equilibrium strategy profile) be found? Our idea is to use the sufficient conditions (Theorem 1) reducing Nash equilibrium strategy profile design to saddle point calculation in a special Germeier convolution of the payoff functions. As an application, this chapter establishes the existence of the Pareto-optimal Nash equilibrium (PoNE) strategy profile in the class of mixed strategies (see Assertion 1). Similar results were obtained by the authors for the Pareto-optimal Berge equilibrium in [4].

Note that two approaches can be adopted to perform formalization of the Pareto unimprovable Nash equilibrium. According to the first approach, Pareto optimality is required on the set of all strategy profiles in the game. The second approach dictates to find the Pareto-optimal equilibrium on the set of all Nash equilibria. Generally, the first approach implies construction of all Nash equilibrium strategy profiles with subsequent check belonging to the Pareto boundary of the strategy profile set of the game (see [5]). Numerical algorithms realizing this approach were suggested for the bimatrix games in [5], for some two-player normal-form games in [6] and the monograph ([7], pp. 92–93), as well as for the linear two-player positional games with cylindrical terminal payoff functions in [8]. In the case of nonlinear differential games with convex terminal payoff functions, the publication [9] obtained the sufficient conditions under which the unimprovable equilibrium strategy profile on the set of Nash equilibria (the second approach) is Pareto optimal on the whole strategy profile set of the game.

This chapter adheres to the second approach, suggesting an algorithm that yields the Pareto-optimal strategy profile among all Nash equilibria.

2. Internally instable set of Nash equilibrium strategy profiles

As is well known, the game theory is used in modeling interactions in economics, sociology, political science, and many other areas. Game theory is the mathematical study of conflict, in which a decision-maker's success in making choices depends on the choice of others. In contrast to the decision-making theory, in game theory, several decision-makers act simultaneously. These decision-makers are called players. Their actions are called pure strategies. Each of the players seeks to achieve their own goals that do not coincide with the goals of other players. A measure of a player's approach of a goal is estimated by his payoff function. The realized value of the player's payoff function is called his payoff. At the same time, the player's payoff function depends not only on his choice but also on the choice of all other players. Therefore, when making a decision, the player is forced to focus not only on his own interests but also on the possible actions of the other players. If the players cannot coordinate their actions, the game is called a noncooperative game. The basic concept of a solution in a noncooperative game theory is the Nash equilibrium.

Consider a noncooperative game (NG) of N players in the class of pure strategies (a non-antagonistic game)

$$\Gamma = \left\langle \mathbb{N}, \{X_i\}_{i\in\mathbb{N}}, \{f_i(x)\}_{i\in\mathbb{N}} \right\rangle, \tag{1}$$

where $\mathbb{N} = \{1, 2, ..., N\}$ is the set of players' serial numbers; each player i chooses and applies his own pure strategy $x_i \in X_i \subseteq \mathrm{R}^{n_i}$, forming no coalition with the others, which induces a strategy profile $x = (x_1, ..., x_N) \in X = \prod_{i\in\mathbb{N}} X_i \subseteq \mathrm{R}^n$ $(n = n_1 + ... + n_N)$; for each $i \in \mathbb{N}$, a payoff function $f_i(x)$ is defined on the strategy profile set X, which gives the payoff of player i. In addition, denote $f = (f_1, ..., f_N)$ and $(x\|z_i) = (x_1, ..., x_{i-1}, z_i, x_{i+1}, ..., x_N)$.

Definition 1. A strategy profile $x^e = (x_1^e, ..., x_N^e) \in X$ is called a Nash equilibrium in the game (1) if

$$\max_{x_i \in X_i} f_i(x^e\|x_i) = f_i(x^e) \quad (i \in \mathbb{N}). \tag{2}$$

The set of all $\{x^e\}$ in game (1) will be designated by X^e.

Now, consider internal instability of X^e. A subset $X^* \subseteq R^n$ is *internally instable* if there exist at least two strategy profiles $x^{(j)} \in X^*$ ($j = 1, 2$) such that

$$\left[f\left(x^{(1)}\right) < f\left(x^{(1)}\right)\right] \Leftrightarrow \left[f_i\left(x^{(1)}\right) < f_i\left(x^{(1)}\right) \ \forall i \in \mathbb{N}\right], \tag{3}$$

internally stable otherwise.

Example 1. Consider a two-player NG of the form

$$\left\langle \{1,2\}, \{X_i = [-1,1]\}_{i=1,2}, \{f_i(x) = -x_i^2 + 2x_1x_2\}_{i=1,2} \right\rangle. \tag{4}$$

A strategy profile $x^e = (x_1^e, x_2^e) \in [-1,1]^2$ is a Nash equilibrium in game (4) if

$$-x_1^2 + 2x_1x_2^e \leq -\left(x_i^e\right)^2 + 2x_1^e x_2^e, \quad -x_2^2 + 2x_1^e x_2 \leq -\left(x_i^e\right)^2 + 2x_1^e x_2^e \quad \forall x_1, x_2 \in [-1,1],$$

which is equivalent to

$$-\left(x_1 - x_2^e\right)^2 \leq -\left(x_1^e - x_2^e\right)^2, \quad -\left(x_1^e - x_2\right)^2 \leq -\left(x_1^e - x_2^e\right)^2.$$

Therefore, we have $X^e = \{(\alpha, \alpha) | \forall \alpha \in [-1,1]\}$ and $f_i(X^e) = \cup_{x^e \in X^e} f_i(x^e) = \cup_{\alpha \in [-1,1]} (\alpha^2, \alpha^2)$ in game (4). Consequently, the set X^e is internally instable in game (4); as for $x^{(1)} = (0,0)$ and $x^{(2)} = (1,1)$, it follows that $f_i(x^{(1)}) = 0 < f_i(x^{(2)}) = 1$ ($i = 1, 2$) (see Eq. (3)).

Note 1. In the antagonistic setting of game (1) ($\mathbb{N} = \{1,2\}$ and $f_1(x) = -f_2(x)$), the equality $f_1(x^{(1)}) = f_1(x^{(2)})$ holds for any two saddle points $x^{(j)} \in X$ ($j = 1, 2$) by the saddle point equivalence. Hence, the saddle point set is always internally stable in the antagonistic game. Note that a saddle point is a Nash equilibrium strategy profile in the antagonistic setting of game (1).

Note 2. In the non-antagonistic setting of game (1), the internal instability effect vanishes if there exist a unique Nash equilibrium strategy profile in (1).

Associate the following auxiliary N-criterion problem with game (1):

$$\Gamma_v = \left\langle X^e, \{f_i(x)\}_{i \in \mathbb{N}} \right\rangle, \tag{5}$$

where the set X^e of *alternatives* x coincides with the set of Nash equilibrium strategy profiles x^e in game (1) and the ith criterion $f_i(x)$ is the payoff function of player i.

Definition 2. An alternative $x^P \in X^e$ is Pareto optimal (efficient) in problem (5) if $\forall x \in X^e$ the system of inequalities

$$f_i(x) \geq f_i\left(x^P\right) \quad (i \in \mathbb{N})$$

is infeasible, with at least one being a strict inequality. Designate by X^P the set of all $\{x^P\}$.

According to Definition 2, the set X^P satisfies the inclusion $X^P \subseteq X^e$ and is *internally stable*.

The following *statement* is obvious: if for all $x \in X^e$ we have

$$\sum_{i \in \mathbb{N}} f_i(x) \leq \sum_{i \in \mathbb{N}} f_i\left(x^P\right), \tag{6}$$

then x^P gives the Pareto-optimal alternative in problem (5).

3. Sufficient conditions of Pareto-optimal equilibrium

Get back to game (1), associating it with the N-criterion problem (5).

Definition 3. A strategy profile $x^* \in X$ is called a Pareto-optimal Nash equilibrium for game (1) if x^* is a Nash equilibrium in (1) (Definition 1) and a Pareto optimum in (5) (Definition 2).

Note 3. Two classes of games where the Pareto equilibrium strategy profiles exist in pure strategies were presented in ([7], pp. 91–92) and, in the case of differential games, in [9–12].

Note 4. Within Example 1, we have two Pareto equilibrium strategy profiles, namely, $x^* = (1, 1)$ and $x^{**} = (-1, -1)$.

Based on (2) and (5), introduce $N+1$ scalar functions defined by

$$\begin{aligned} \varphi_i(x,z) &= f_i(z\|x_i) - f_i(z) \ (i \in \mathbb{N}), \\ \varphi_{N+1}(x,z) &= \sum_{r\in\mathbb{N}} f_r(x) - \sum_{r\in\mathbb{N}} f_r(z), \end{aligned} \tag{7}$$

where $z = (z_1, ..., z_N)$, $z_i \in X_i$ $(i \in \mathbb{N})$, $z \in X$, $x \in X$. The Germeier convolution ([13], p. 43) of the scalar functions (7) has the form

$$\varphi(x,z) = \max_{j=1,\dots,N+1} \varphi_j(x,z). \tag{8}$$

In addition, associate the following *antagonistic* game with game (1) and the N-criterion problem (5):

$$\langle X, Z = X, \varphi(x, \mathrm{z})\rangle. \tag{9}$$

In this game, player 1 and his opponent choose their strategies $x \in X$ and $z \in X$ to maximize and minimize, respectively, the payoff function $\varphi(x,z)$ described by (7) and (8).

A saddle point $(x^o, z^*) \in X^2$ of game (9) is defined by the chain of inequalities

$$\varphi(x,z^*) \le \varphi(x^0,z^*) \le \varphi(x^0,z) \quad \forall x, z \in X. \tag{10}$$

In game (9), the saddle points are given by the minimax strategy z^*

$$\left(\min_{z\in X}\max_{x\in X}\varphi(x,\mathrm{z}) = \max_{x\in X}\varphi(x,\mathrm{z}^*)\right)$$

and the maximin strategy x^0

$$\left(\max_{x\in X}\min_{z\in X}\varphi(x,\mathrm{z}) = \min_{z\in X}\varphi(x^0,\mathrm{z})\right).$$

The following statement defines *a sufficient condition* for the existence of a PoNE strategy profile in game (1).

Theorem 1. If a saddle point (x^o, z^*) exists in the antagonistic game (9) (i.e., the condition (10) holds), then the minimax strategy z^* is a PoNE strategy profile for game (1) [14].

Proof. Let $z = x^0$ for the right-hand inequality in (10). Using (7) and (8), we have

$$\varphi(x^0,x^0) = \max_{j=1,\dots,N+1} \varphi_j(x^0,x^0) = 0.$$

By (10), for all $x \in X$ it follows that

$$0 \geq \varphi(x, z^*) = \max_{j=1,\dots,N+1} \varphi_j(x, z^*) = 0.$$

Therefore, for all $x \in X$, the following chain of implications is true:

$$\left[0 \geq \max_{j=1,\dots,N+1} \varphi_j(x, z^*) \geq \varphi_j(x, z^*)\right] \Rightarrow$$

$$\Rightarrow \left[\varphi_j(x, z^*) \leq 0 \ (j = 1, \dots, N, N+1)\right] \overset{(7)}{\Rightarrow}$$

$$\overset{(7)}{\Rightarrow} \left\{ \left[f_j(z^* \| x_i) - f_j(z^*) \leq 0 \ \forall x_i \in X_i \ \ (i \in \mathbb{N})\right] \wedge \right.$$

$$\left. \wedge \left[\sum_{r \in \mathbb{N}} f_r(x) - \sum_{r \in \mathbb{N}} f_r(z^*) \leq 0 \ \forall x \in X^e\right] \right\} \Rightarrow$$

$$\Rightarrow \left\{ \left[\max_{x_i \in X_i} f_j(z^* \| x_i) = f_j(z^*) \ (i \in \mathbb{N})\right] \wedge \right.$$

$$\left. \wedge \left[\max_{x \in X^e} \sum_{i \in \mathbb{N}} f_i(x) = \sum_{i \in \mathbb{N}} f_i(z^*)\right] \right\} \overset{(2),(6)}{\Rightarrow} \left\{ [z^* \in X^e] \wedge [z^* \in X^P] \right\}.$$

This chain involves the inclusion $X^e \subseteq X$. □

Remark 1. Theorem 1 substantiates the following design method of the PoNE strategy profile x^* in game (1).

Step 1. Using the payoff functions $f_i(x)$ $(i \in \mathbb{N})$ from (1) and the vectors $z = (z_1, \dots, z_N)$, $z_i \in X_i$ and $x = (x_1, \dots, x_N)$, $x_i \in X_i$ $(i \in \mathbb{N})$, construct the function $\varphi(x, z)$ by formulas (7) and (8).

Step 2. Find the saddle point $(x^o, , z^*)$ of antagonistic game (9). Then z^* is the Pareto equilibrium solution of game (1).

As far as the authors know, numerical calculation methods of the saddle point (x^o, z^*) for the Germeier convolution

$$\varphi(x, z) = \max_{j=1,\dots,N+1} \varphi_j(x, z)$$

have not been developed yet. However, they are vital to construct the Nash equilibrium strategy profiles that are Pareto optimal (see Theorem 1). This is a new trend in equilibrium programming; in the authors' opinion, it can be developed using the mathematical apparatus of Germeier convolution optimization $\max_j \varphi_j(x)$ proposed by Dem'yanov [15].

Remark 2. The results of operations research ([16], p. 54) yield the following statement that is crucial to prove the existence of a PoNE strategy profile in the class of mixed strategies in game (1) (see the forthcoming section). If $X_i \in comp\mathbb{R}^{n_i}$ and $f_i(\cdot) \in C(X)(i \in \mathbb{N})$ in game (1), then the Germeier convolution $\varphi(x, z) = \max_{j=1,\dots,N+1} \varphi_j(x, z)$ from (7) and (8) is continuous on $X \times X$.

4. Existence of PoNE strategy profile in mixed strategies

That game (1) admits a PoNE strategy profile in the class of pure strategies (see Definition 3) is rather a miracle. This equilibrium may exist only for special payoff functions, strategy sets, and numbers of players. Therefore, adhering to the approach associated with E. Borel [17], J. von Neumann [18], Nash [1], and their followers, we establish the existence of the PoNE strategy profile of game (1) in the class of mixed strategies under standard game theory restrictions (i.e., compact strategy sets and continuous payoff functions).

And so, suppose that in game (1) the sets X_i of the pure strategies x_i are compact sets in R^{n_i} (are closed and bounded), whereas the payoff function $f_i(x)$ of each player i ($i \in \mathbb{N}$) is continuous on the set of pure strategy profiles X.

Consider the *mixed strategy extension of game* (1). To this end, construct the Borel σ-algebra $\mathfrak{B}(X_i)$ on each compact set X_i ($i \in \mathbb{N}$) and probability measures $\nu_i(\cdot)$ on $\mathfrak{B}(X_i)$ (i.e., nonnegative scalar functions defined on the elements of $\mathfrak{B}(X_i)$ that are countably additive and normalized to unity on X_i). Denote by $\{\nu_i\}$ the whole set of such measures; the measure $\nu_i(\cdot)$ proper is called the *mixed strategy of player i* ($i \in \mathbb{N}$) in game (1). Next, for game (1) construct the *mixed strategy profiles*, that is, the multiplicative measures

$$\nu(dx) = \nu_1(dx_1)...\nu_N(dx_N),$$

and designate by $\{\nu\}$ the set of such strategy profiles. And finally, find the mathematical expectations

$$f_i(\nu) = \int_X f_i(x)\nu(dx) \ (i \in \mathbb{N}). \tag{11}$$

As a result, the game Γ from (1) is associated with its *mixed strategy extension*

$$\tilde{\Gamma} = \left\langle \mathbb{N}, \{\nu_i\}_{i \in \mathbb{N}}, \{f_i(\nu)\}_{i \in \mathbb{N}} \right\rangle.$$

In the noncooperative game $\tilde{\Gamma}$, we have the following elements:

$\nu_i(\cdot) \in \{\nu_i\}$ as the mixed strategy of player i.

$\nu(\cdot) \in \{\nu\}$ as the mixed strategy profile.

$f_i(\nu)$ as the payoff function of player i defined by (11).

Further exposition involves the vector $z = (z_1, ..., z_N) \in X$ with $z_i \in X_i$ ($i \in \mathbb{N}$), and, of course, the vector $x = (x_1, ..., x_N) \in X,$ as well as the mixed strategy profiles $\nu(\cdot), \mu(\cdot) \in \{\nu\}$ and the mathematical expectations

$$f_i(\nu) = \int_X f_i(x)\nu(dx), \quad f_i(\mu) = \int_X f_i(z)\mu(dz),$$

$$f_i(\mu\|\nu_i) = \int_{X_1} \cdots \int_{X_{i-1}} \int_{X_i} \int_{X_{i+1}} \cdots \int_{X_N} f_i(x)\mu_N(dz_N)... \tag{12}$$

$$...\mu_{i+1}(dz_{i+1})\nu_i(dx_i)\mu_{i-1}(dz_{i-1})...\mu_1(dz_1).$$

Once again, we underline that $x_i, z_i \in X_i$ ($i \in \mathbb{N}$) and $x, z \in X$.

The following notion of the Nash equilibrium strategy profile $\nu^e(\cdot) \in \{\nu\}$ in mixed strategies in original game (1) answers to Definition 1 of the Nash equilibrium strategy profile $x^e \in X$ in pure strategies in the same game (1).

Definition 4. A strategy profile $\nu^e(\cdot) \in \{\nu\}$ is called a Nash equilibrium for the game $\tilde{\Gamma}$ if

$$f_i(\nu^e\|\nu_i) \leq f_i(\nu^e) \ \forall \nu_i(\cdot) \in \{\nu_i\} \ (i \in \mathbb{N}); \tag{13}$$

throughout the paper, $\nu^e(\cdot) \in \{\nu\}$ will be also called the Nash equilibrium strategy profile in mixed strategies for game (1).

By the Glicksberg theorem [19], there exists a Nash equilibrium strategy profile in mixed strategies in game (1) under $X_i \in comp\mathrm{R}^{n_i}$ and $f_i(\cdot) \in C(X) (i \in \mathbb{N})$. Denote by $\mathfrak{N}$ the set of such profiles $\{\nu^e\}$.

Associate the following N-criterion problem with the game $\tilde{\Gamma}$

$$\tilde{\Gamma}_\nu = \left\langle \mathfrak{N}, \{f_i(\nu)\}_{i \in \mathbb{N}} \right\rangle. \tag{14}$$

In (14), a decision-maker chooses a strategy profile $\nu(\cdot) \in \mathfrak{N}$ to simultaneously maximize all components of the vector criterion $f(\nu) = (f_1(\nu), \ldots, f_N(\nu))$. The notion of the Pareto optimal strategy profile is conventional (see below).

Definition 5. A strategy profile $\nu^P(\cdot) \in \mathfrak{N}$ is called Pareto optimal for the N-criterion problem $\tilde{\Gamma}_\nu$ from (14) if for any $\nu(\cdot) \in \mathfrak{N}$ the system of inequalities

$$f_i(\nu) \geq f_i(\nu^P) \ (i \in \mathbb{N})$$

is infeasible, with at least one inequality being strict.

The following **statement** represents an analog of (6): if for all $\nu(\cdot) \in \mathfrak{N}$ we have

$$\sum_{i \in \mathbb{N}} f_i(\nu) \leq \sum_{i \in \mathbb{N}} f_i(\nu^P), \tag{15}$$

then the mixed strategy profile $\nu^P(\cdot) \in \mathfrak{N}$ is Pareto optimal in the problem $\tilde{\Gamma}_\nu$ from (14).

Combining Definition 4 with Definition 5 leads to.

Definition 6. A strategy profile $\nu^*(\cdot) \in \{\nu\}$ is called a Pareto-optimal Nash equilibrium strategy profile in mixed strategies for game (1) if $\nu^*(\cdot)$ is a Nash equilibrium in $\tilde{\Gamma}$ (according to Definition 4), and $\nu^*(\cdot)$ is Pareto optimal in the multicriterion problem $\tilde{\Gamma}_\nu$ (according to Definition 5).

Now, we prove the existence of a Nash equilibrium strategy profile in mixed strategies that is Pareto optimal with respect to the rest Nash equilibrium strategy profiles.

Assertion 1. Consider the noncooperative game (1) where:

1. The pure strategy set X_i of each player i is a nonempty compact set in R^{n_i} $(i \in \mathbb{N})$.

2. The payoff function $f_i(x)$ of player i $(i \in \mathbb{N})$ is continuous on the strategy profile set X.

Then there exists a PoNE strategy profile in mixed strategies in game (1).

Proof. Using formulas (7) and (8), construct the scalar function

$$\varphi(x, z) = \max_{j=1, \ldots, N+1} \varphi_j(x, z),$$

where

$$\varphi_i(x, z) = f_i(z \| x_i) - f_i(z) \quad (i \in \mathbb{N}),$$
$$\varphi_{N+1}(x, z) = \sum_{r \in \mathbb{N}} f_r(x) - \sum_{r \in \mathbb{N}} f_r(z),$$

According to the construction procedure and Remark 2, the function $\varphi(x,z)$ is defined and continuous on the product of compact sets $X \times X$.

Define the auxiliary antagonistic game

$$\Gamma_a = \langle \{\mathrm{I}, \mathrm{II}\}, X, Z = X, \varphi(x, z) \rangle,$$

where players I and II seek to maximize and minimize, respectively, the function $\varphi(x,z)$ continuous on $X \times Z(Z = X)$ by choosing their strategies $x \in X$ and $z \in X$.

Now, apply a special case of the Glicksberg theorem [19] to the game Γ_a, as the saddle point in this game coincides with the Nash equilibrium strategy profile in the two-player noncooperative game

$$\Gamma_2 = \left\langle \{\mathrm{I}, \mathrm{II}\}, \{X, Z = X\}, \{f_{\mathrm{I}}(x, z) = \varphi(x, z), f_{\mathrm{II}}(x, z) = -\varphi(x, z)\} \right\rangle.$$

In this game, player I seeks to maximize $f_{\mathrm{I}}(x, z) = \varphi(x, z)$ by choosing his strategy $x \in X$, whereas player II tries to maximize $f_{\mathrm{II}}(x, z) = -\varphi(x, z)$ The sets X and $X = Z$ in game Γ_2 are compact, while the payoff functions $f_{\mathrm{I}}(x, z)$ and $f_{\mathrm{II}}(x, z)$ are continuous on $X \times Z$; hence, by the Glicksberg theorem, there exists a Nash equilibrium strategy profile (ν^e, μ^*) in the mixed extension Γ_2:

$$\tilde{\Gamma}_2 = \left\langle \{\mathrm{I}, \mathrm{II}\}, \{\nu\}, \{\mu\}, \left\{ f_i(\nu, \mu) = \int_X \int_X f_i(x,z)\nu(dx)\mu(dz) \right\}_{i=\mathrm{I},\,\mathrm{II}} \right\rangle.$$

In addition, (ν^e, μ^*) is simultaneously a saddle point of the mixed extension of the game Γ_a :

$$\tilde{\Gamma}_a = \left\langle \{\mathrm{I}, \mathrm{II}\}, \{\nu\}, \{\mu\}, \varphi(\nu, \mu) = \int_X \int_X \varphi(x,z)\nu(dx)\mu(dz) \right\rangle.$$

Thus, according to the Glicksberg theorem, there exists a pair (ν^e, μ^*) representing a saddle point of $\varphi(\nu, \mu)$, that is,

$$\varphi(\nu, \mu^*) \leq \varphi(\nu^e, \mu^*) \leq \varphi(\nu^e, \mu), \quad \forall \nu(\cdot), \mu(\cdot) \in \{\nu\}. \tag{16}$$

Letting $\mu = \nu^e$ in the right inequality of (16) gives $\varphi(\nu^e, \nu^e) = 0$ and so, $\forall \nu(\cdot) \in \{\nu\}$ formula (16) implies

$$0 \geq \varphi(\nu, \mu^*) = \int_X \int_X \max_{j=1,\dots,N+1} \varphi_j(x,z)\nu(dx)\mu^*(dz). \tag{17}$$

It was established in [3] that

$$\max_{j=1,\dots,N+1} \int_X \int_X \varphi_j(x,z)\nu(dx)\mu(dz) \leq \int_X \int_X \max_{j=1,\dots,N+1} \varphi_j(x,z)\nu(dx)\mu(dz). \tag{18}$$

Note that this property has an analog: the maximum of the sum of functions does not exceed the sum of their maxima. It follows from (17) and (18) that

$$\max_{j=1,\dots,N+1} \int_X \int_X \varphi_j(x,z)\nu(dx)\mu^*(dz) \leq 0 \ \forall \nu(\cdot) \in \{\nu\},$$

and then surely for each $j = 1,\dots, N, N+1$, we have

$$\int_X \int_X \varphi_j(x,z)\nu(dx)\mu^*(dz) \le 0 \ \forall \nu(\cdot) \in \{\nu\}. \tag{19}$$

Next, taking into account the normalized mixed strategies and the normalized mixed strategy profiles, that is, the conditions

$$\int_X \nu_i(dx_i) = 1, \quad \int_X \mu_i(dz_i) = 1 (i \in \mathbb{N}), \quad \int_X \nu(dx) = 1, \quad \int_X \mu(dz) = 1 \tag{20}$$

that hold $\forall \nu_i(\cdot) \in \{\nu_i\}, \mu_i(\cdot) \in \{\mu_i\}, \nu(\cdot) \in \{\nu\}, \mu(\cdot) \in \{\mu\}$, we distinguish between two cases, namely, $j \in \mathbb{N}$ and $j = N+1$. For each of these cases, it is necessary to refine inequalities (19).

Case 1: $j \in \mathbb{N}$. Using (7) and (20) for each $i \in \mathbb{N}$, inequality (19) is reduced to the form

$$\int_X \int_X [f_i(z\|x_i) - f_i(z)]\nu(dx)\mu^*(dz) = \int_X \int_{X_i} [f_i(z\|x_i) - f_i(z)]\nu_i(dx_i)\mu^*(dz) =$$

$$= \int_X \int_X f_i(z\|x_i)\nu_i(dx_i)\mu^*(dz) - \int_X f_i(z)\mu^*(dz) \int_{X_i} \nu_i(dx_i) \overset{(12),(20)}{=}$$

$$\overset{(12),(20)}{=} \left[\int_{X_1} \dots \int_{X_{i-1}} \int_{X_i} \int_{X_{i+1}} \dots \int_{X_N} f_i(z_1, \dots, z_{i-1}, x_i, z_{i+1}, \dots, z_N)\mu_N^*(dz_N)\dots \right.$$

$$\left. \dots\mu_{i+1}^*(dz_{i+1})\nu_i(dx_i)\mu_{i-1}^*(dz_{i-1})\dots\mu_1^*(dz_1) \right] - f_i(\mu^*) =$$

$$= f_i(\mu^*\|\nu_i) - f_i(\mu^*) \le 0 \ \forall \nu_i(\cdot) \in \{\nu_i\}.$$

In combination with (13), this result gives the inclusion $\mu^*(\cdot) \in \mathfrak{N}$, that is, the mixed strategy profile $\mu^*(\cdot)$ is a Nash equilibrium for the game (1) by Definition 4.

Case 2: $j = N+1$. Here inequality (19) acquires the form

$$\int_X \int_X \varphi_{N+1}(x,z)\nu(dx)\mu^*(dz) \overset{(7)}{=} \int_X \int_X \sum_{i \in \mathbb{N}} f_i(x)\nu(dx)\mu^*(dz) - \int_X \int_X \sum_{i \in \mathbb{N}} f_i(x)\nu(dx)\mu^*(dz) =$$

$$= \int_X \sum_{i \in \mathbb{N}} f_i(x)\nu(dx) \int_X \mu^*(dz) - \int_X \sum_{i \in \mathbb{N}} f_i(z)\mu^*(dz) \int_X \nu(dx) \overset{(20)}{=}$$

$$\overset{(20)}{=} \sum_{i \in \mathbb{N}} \int_X f_i(x)\nu(dx) - \sum_{i \in \mathbb{N}} \int_X f_i(z)\mu^*(dz) \overset{(12)}{=} \sum_{i \in \mathbb{N}} f_i(\nu) - \sum_{i \in \mathbb{N}} f_i(\mu^*) \le 0 \ \forall \nu(\cdot) \in \mathfrak{N},$$

in as much as $\mathfrak{N} \subseteq \{\nu\}$. This immediately yields (15) for $\nu^P = \mu^*$, that is, the strategy profile $\mu^*(\cdot)$ is Pareto optimal for the N-criterion problem $\tilde{\Gamma}_\upsilon$ from (14) by Definition 5.

This outcome and the inclusion $\mu^*(\cdot) \in \mathfrak{N}$ conclude the proof. □

Note 5. Another proof of Assertion 1 can be found in ([3], pp. 13–15).

5. Conclusions

Vorob'ev, the founder of game theory in Russia, believed that its subject [20] is answering the following three questions:

1. What is the optimality of a given game?

2. Does an optimal solution exist?

3. How can it be found?

For the many-player noncooperative games, the answer to the first question is the PoNE strategy profile.

The answer to the second question is given by Assertion 1: if the strategy sets are compact and the payoff functions are continuous, then a Pareto equilibrium strategy profile exists in the class of mixed strategies.

As turned out, the answer to the third question is not so simple. At first glance, one should just construct the Germeier convolution of the payoff functions using formulas (7) and (8) and find the saddle point (10); then the minimax strategy entering the saddle point is the PoNE strategy profile. This equilibrium design method is dictated by Theorem 1, actually being the basic result of the present paper. However, the issues of saddle point construction for the Germeier convolutions have not been developed so far. The usage of specific numerical algorithms and their complexity still remain under investigated. Further research by the authors and, hopefully, by the readers will endeavor to improve the situation.

Author details

Vladislav Zhukovskiy[1] and Konstantin Kudryavtsev[2*]

1 M.V. Lomonosov Moscow State University, Moscow, Russia

2 South Ural State University, Chelyabinsk, Russia

*Address all correspondence to: kudrkn@gmail.com

References

[1] Nash JF. Equilibrium points in N-Person games. Proceedings of the National Academy of Sciences of the United States of America. 1950;**36**:48-49

[2] Zhukovskiy VI, Kudryavtsev KN. Equilibrating conflicts under uncertainty. I. Analogue of a saddle-point. Matematicheskaya Teoriya Igr i Ee Prilozheniya. 2013;**5**(1):27-44

[3] Zhukovskiy VI, Kudryavtsev KN. Equilibrating Conflicts under Uncertainty. II. Analogue of a Maximin. Matematicheskaya Teoriya Igr i Ee Prilozheniya. 2013;**5**(2):3-45

[4] Zhukovskiy VI, Kudryavtsev KN. Mathematical foundations of the Golden Rule. I. Static case. Automation and Remote Control. 2017;**78**(10): 1920-1940. DOI: 10.1134/S0005117917100149

[5] Gatti N, Rocco M, Sandhom T. On the verification and computation of strong Nash equilibrium. In: Proceedings of the ACM International Joint Conference on Autonomous Agents and Multi-Agent Systems (AAMAS); 6–10 May 2013; Saint Paul, USA. 2013. pp. 723-730

[6] Gasko N, Suciu M, Lung RI, Dumitrescu D. Pareto-optimal Nash equilibrium detection using an evolutionary approach. Acta Universitatis Sapientiae. 2012;**4**(2): 237-246

[7] Podinovskii VV, Nogin BD. Pareto Optimal Solutions of Multicriteria Problems. Moscow: Fizmatlit; 2007

[8] Kleimenov AF, Kuvshinov DR, Osipov SI. Nash and Stackelberg solutions numerical construction in a two-person nonantagonistic linear positional differential game. Trudy Instituta Matematiki i Mekhaniki UrO RAN. 2009;**15**(4):120-133

[9] Mamedov MB. Investigation of unimprovable equilibrium situations in nonlinear conflict-controlled dynamical systems. Zhurnal Vychislitel'noy Matematiki i Matematicheskoy Fiziki. 2004;**44**(2):308-317

[10] Kononenko AF, Konurbaev EM. Existence of equilibrium profiles in the class of positional strategies that are pareto optimal for certain differential games. In: Game Theory and Its Applications. Kemerovo: Kemerovo State University; 1983. pp. 105-115

[11] Mamedov MB. On a Nash Equilibrium Profile Being Pareto Optimal. Izvestiya Akademii Nauk Azerbaĭdzhana: Seriya Fiziko-Tekhnicheskikh. 1983;**4**(2):11-17

[12] Starr AW, Ho YC. Further properties of nonzero-sum differential games. The Journal of Optimization Theory and Applications. 1969;**3**(4): 207-219

[13] Germeier YB. Introduction to the Theory of Operations Research. Moscow: Nauka; 1971

[14] Kudryavtsev K, Zhukovskiy V, Stabulit I. One method for computing the Pareto-optimal Nash equilibrium in bimatrix game. In: 2017 Constructive Nonsmooth Analysis and Related Topics (dedicated to the memory of VF Demyanov)(CNSA). IEEE; 2017. pp. 1-3. DOI: 10.1109/CNSA.2017.7973978

[15] Dem'yanov VF, Malozemov VN. Introduction to Minimax. Moscow: Nauka; 1972

[16] Morozov VV, Sukharev AG, Fedorov VV. Operations Research in Problems and Exercises. Moscow: Nauka; 1986

[17] Borel E. La th'eorie du jeu et les equations int'engrales a noyau

sym'etrique. Comptes rendus de l'Académie des Sciences. 1921;**173**: 1304-1308

[18] von Neumann J. Zur Theorie der Gesellschaftspiele. Mathematische Annalen. 1928;**100**(1):295-320

[19] Glicksberg IL. A further generalization of the Kakutani fixed point theorem, with application to Nash equilibrium points. Proceedings of American Mathematical Society. 1952; **3**(1):170-174

[20] Vorob'ev NN. The present state of the theory of games. Russian Mathematical Surveys. 1970;**25**(2): 77-136

Chapter 2

Polynomial Algorithm for Constructing Pareto-Optimal Schedules for Problem $1|r_j|L_{\max}, C_{\max}$

Alexander A. Lazarev and Nikolay Pravdivets

Abstract

In this chapter, we consider the single machine scheduling problem with given release dates, processing times, and due dates with two objective functions. The first one is to minimize the maximum lateness, that is, maximum difference between each job due date and its actual completion time. The second one is to minimize the maximum completion time, that is, to complete all the jobs as soon as possible. The problem is NP-hard in the strong sense. We provide a polynomial time algorithm for constructing a Pareto-optimal set of schedules on criteria of maximum lateness and maximum completion time, that is, problem $1|r_j|L_{max}, C_{max}$, for the subcase of the problem: $d_1 \leq d_2 \leq ... \leq d_n; d_1 - r_1 - p_1 \geq d_2 - r_2 - p_2 \geq ... \geq d_n - r_n - p_n$.

Keywords: single machine scheduling, two-criteria scheduling, Pareto-set, Pareto-optimality, minimization of maximum lateness, minimization of maximum completion time, polynomial time algorithm

1. Introduction

We consider a classical scheduling problem on a single machine. A release time of each job is predefined and represents the minimum possible start time of the job. When constructing schedules, we consider two objective functions. The first one is to minimize the maximum lateness, that is, maximum difference between each job due date and its actual completion time. The second one is to minimize the maximum completion time, that is, to complete all the jobs as soon as possible. The problem is NP-hard in the strong sense [1]. We provide a polynomial time algorithm for constructing a Pareto-optimal set of schedules on criteria of maximum lateness and maximum completion time, that is, problem $1|r_j|L_{max}, C_{max}$, for the subcase of the problem when due dates are: $d_1 \leq d_2 \leq ... \leq d_n;\ d_1 - r_1 - p_1 \geq d_2 - r_2 - p_2 \geq ... \geq d_n - r_n - p_n$. Example of a problem case that meets these constraints will be the case when all jobs have the same time for processing before due date.

2. Statement of the problem $1|r_j|L_{\max}, C_{\max}$

We consider single machine scheduling problem, where a set of n jobs $N = \{1, 2, ..., n\}$ has to be processed on a single machine. Each job we is numbered, that is, the entry "job j" is equivalent to the entry "job numbered j." Simultaneous executing of jobs or preemptions of the processing of a job are prohibited. For jobs $j \in N$, value r_j is the minimum possible start time, $p_j \geq 0$ is a processing time of job j and d_j is a due date of job j.

The schedule is represented by a set $\pi = \{s_j | j \in N\}$ of start times of each job. By τ, we denote the permutation of $(j_1, \ldots, j_n)$ elements of the set N. A set of all different schedules of jobs from the set N is denoted by $\Pi(N)$. Schedule π is called *feasible* if $s_j(\pi) \geq r_j$, $\forall j \in N$. We denote the completion time of job $j \in N$ in schedule π as $C_j(\pi)$. Difference $L_j(\pi) = C_j(\pi) - d_j, j \in N$, denotes lateness of job j in the schedule π. Maximum lateness of jobs of the set N under the schedule π is

$$L_{\max}(\pi) = \max_{j \in N} \{C_j(\pi) - d_j\}. \tag{1}$$

We denote the completion time of all jobs of the set N in schedule π as

$$C_{\max}(\pi) = \max_{j \in N} C_j(\pi).$$

The problem is to find the optimal schedule π^* with the smallest value of the maximum lateness:

$$L^*_{\max} = \min_{\pi \in \Pi(N)} L_{\max}(\pi) = L_{\max}(\pi^*). \tag{2}$$

For any arbitrary set of jobs $M \subseteq N$ we also denote:

$$r_M = \min_{j \in M} r_j, \quad d_M = \max_{j \in M} d_j, \quad p_M = \sum_{j \in M} p_j. \tag{3}$$

In the standard notation of Graham et al. [2], this problem is denoted as $1|r_j|L_{\max}$. Intensive work on the solution of this problem has continued since the early 50s of the 20th century. Lenstra et al. [1] showed that the general case of the problem $1|r_j|L_{\max}$ is *NP*-hard in the strong sense.

Potts [3] introduced an iterative version of extended Jackson rule (IJ) [4] and proved that $\frac{L_{\max}(\pi_{IJ})}{L^*_{\max}} \leq \frac{3}{2}$. Hall and Shmoys [5] modified the iterative version and created an algorithm (MIJ) that guarantees the evaluation $\frac{L_{\max}(\pi_{MIJ})}{L^*_{\max}} \leq \frac{4}{3}$. They also presented two approximation schemes that guarantee finding ε-approximate solution in $O\left(n \log n + n(1/\varepsilon)^{O(1/\varepsilon^2)}\right)$ and $O\left((n/\varepsilon)^{O(1/\varepsilon)}\right)$ operations. Mastrolilli [6] introduced an improved approximation scheme with complexity of $O\left(n + (1/\varepsilon)^{O(1/\varepsilon)}\right)$ operations.

A number of polynomially solvable cases of the problem were found, starting with Jackson's early result [4] for the case $r_j = 0, j \in N$, when the solution is a schedule in which jobs are ordered by nondecreasing due dates (by rule *EDD*). Such a schedule is also be optimal for the case when the release times and due dates are associated $(r_i \leq r_j \Leftrightarrow d_i \leq d_j, \forall i, j \in N)$.

Schedule is constructed according to the extended Jackson rule (Schrage schedule): on the next place in the schedule we select a released non-ordered job with the minimum due date; if there are no such jobs, then we select the job with the minimum release time among the unscheduled jobs.

If process times of all jobs are equal, the complexity can be reduced to $O(n \log n)$ [7]. Vakhania generalized this result [8] considering the case when the processing times of some jobs are restricted to either p or $2p$. An algorithm with complexity of $O(n^2 \log n \log p)$ was suggested.

A case when job processing times are mutually divisible is considered in [9]. Author suggest a polynomial-time algorithm with a complexity of $O(n^3 \log n \log^2 p_{\max})$ operations for solving this case.

Special cases $1|prec; r_j|C_{\max}$, $1|prec; p_j = p; r_j|L_{\max}$ and $1|prec; r_j; pmtn|L_{\max}$ with precedence constraints for jobs have been addressed in works of Lawler [10], Simons [11], Baker et al. [12]. Hoogeveen [13] proposed a polynomial algorithm for the special case when job parameters satisfy the constraints $d_j - p_j - A \le r_j \le d_j - A$, $\forall j \in N$, for some constant A. A pseudo-polynomial algorithm for the NP-hard case when release times and due dates are in the reversed order ($d_1 \le \dots \le d_n$ and $r_1 \ge \dots \ge r_n$) was developed in [14].

We denote by $L_j^A(\pi)$ and $C_j^A(\pi)$ the lateness and completion time of job $j \in N$ in schedule π, for instance, A with job parameters $\left\{r_j^A, p_j^A, d_j^A\right\}, j \in N$. Respectively, $L_{\max}^A(\pi) = \max_{j \in N} L_j^A(\pi)$ is a maximum lateness of the schedule π for instance A.

This paper deals with a problem with two objective functions $L_{\max}$ and $C_{\max}$, which in general case can be referred as $1|r_j|L_{\max}, C_{\max}$. This problem was considered in [15], where authors consider some dominance properties and conditions when the Pareto-optimal set can be formed in polynomial time.

Definition 1.1 For any instance A of the problem, each permutation τ of the jobs of the set N is uniquely defines *early schedule* π_τ^A. In the early schedule, each job $j \in N$ starts immediately after the end of the previous job in the corresponding permutation. If the completion time of the previous job is less than the release time of the current job, then the beginning of the current job is equal to its release time. That is, if $\tau = (j_1, j_2, \dots, j_n)$, then $\pi_\tau^A = (s_{j_1}, s_{j_2}, \dots, s_{j_n})$, where

$$s_{j_1} = r_{j_1}^A, s_{j_k} = \max\left\{s_{j_{k-1}} + p_{j_{k-1}}^A, r_{j_k}^A\right\}, \quad k = 2, \dots, n. \tag{4}$$

Early schedules play an important role in our construction, since it is sufficient to check all early schedules to find the optimal schedule of any problem instance.

By τ^A we denote the optimal permutation and π^A we denote the optimal schedule for instance A. Only early optimal schedules are be considered, that is, $\pi^A = \pi_{\tau^A}^A$.

We denote by $\Pi(N)$ the set of all permutations of jobs of the set N, and by Π_A the set of feasible schedules for instance A.

3. Problem $1|d_i \le d_j, d_i - r_i - p_i \ge d_j - r_j - p_j|L_{\max}, C_{\max}$

This section deals with the problem of constructing a Pareto-optimal set by criteria $C_{\max}$ and $L_{\max}$, that is, problem $1|r_j|L_{\max}, C_{\max}$. We suggest an algorithm for constructing a set of schedules $\Phi(N, t) = \left\{\pi_1', \pi_2', \dots, \pi_m'\right\}$ for which

$$C_{\max}(\pi_1') < C_{\max}(\pi_2') < \dots < C_{\max}(\pi_m'), \tag{5}$$

$$L_{\max}(\pi_1') > L_{\max}(\pi_2') > \dots > L_{\max}(\pi_m'). \tag{6}$$

There is no schedule π such that $C_{\max}(\pi) \le C_{\max}(\pi_i')$ and $L_{\max}(\pi) \le L_{\max}(\pi_i')$, at least one of the inequalities is strict for some i, $i = 1, \ldots, m$. It is shown that $m \le n$.

3.1 Problem properties

We denote the precedence of the jobs i and j in schedule π as $(i \to j)_\pi$. We also introduce

$$r_j(t) = \max\{r_j, t\}; \tag{7}$$

$$r(N,t) = \min_{j \in N}\{r_j(t)\}. \tag{8}$$

In cases when its obvious how many jobs we mean, we write $r(t)$ instead of $r(N,t)$.

We assume that the job parameters satisfy the following constraints:

$$d_1 \le \ldots \le d_n, \quad d_1 - r_1 - p_1 \ge \ldots \ge d_n - r_n - p_n. \tag{9}$$

For example, these constraints correspond to the case when $d_j = r_j + p_j + z$, $j = 1, \ldots, n$, where z is a constant, that is, when all jobs have the same time for processing before due date. A problem with similar constraints but for a single objective function ($L_{\max}$) is considered in [16].

We assume that $|N| > 1$ and t is the time when the machine is ready. From the set N, we find two jobs $f = f(N,t)$ and $s = s(N,t)$ in the following way:

$$f(N,t) = \arg\min_{j \in N}\{d_j | r_j(t) = r(N,t)\}, \tag{10}$$

$$s(N,t) = \arg\min_{j \in N \setminus \{f\}}\{d_j | r_j(t) = r(N \setminus f, t)\}, \tag{11}$$

where $f = f(N,t)$. If $N = \{i\}$, then we set $f(N,t) = i$, $s(N,t) = 0$, $\forall t$. We also define $d_0 = +\infty$, $f(\emptyset, t) = 0$, $s(\emptyset, t) = 0$, $\forall t$. For jobs f and s the following properties are true.

Lemma 1.1 If the jobs of the set N satisfy (4), then for any schedule $\pi \in \Pi(N)$ for all $j \in N \setminus \{f\}$, for which $(j \to f)_\pi$

$$L_j(\pi) < L_f(\pi) \tag{12}$$

is true, and for all $j \in N \setminus \{f, s\}$, satisfying the condition $(j \to s)_\pi$,

$$L_j(\pi) < L_s(\pi), \tag{13}$$

where $f = f(N,t)$ and $s = s(N,t)$, is also true.

Proof: For each job j such that $(j \to f)_\pi$, completion time $C_j(\pi) < C_f(\pi)$. If $d_j \ge d_f$, then obviously

$$L_j(\pi) = C_j(\pi) - d_j < C_f(\pi) - d_f = L_f(\pi), \tag{14}$$

therefore (12) is valid.

If for job $j \in N$, $(j \to f)_\pi$, then $d_j < d_f$. Then $r_j > r_f$. If $r_j \le r_f$, then $r_j(t) \le r_f(t)$ and $r_f(t) = r(t)$, as follows from (7) and (10). Then $r_j(t) = r_f(t) = r(t)$ and $d_j < d_f$, but this contradicts the definition of job f (10). Therefore, $r_j > r_f$. Its obvious that $C_j(\pi) - p_j < C_f(\pi) - p_f$ and, since $r_j > r_f$,

$$C_j(\pi) - p_j - r_j < C_f(\pi) - p_f - r_f, \tag{15}$$

$$C_j(\pi) - C_f(\pi) < p_j + r_j - p_f - r_f. \tag{16}$$

Since $d_j < d_f$, then (from (9)) $d_j - r_j - p_j \geq d_f - r_f - p_f$ or $d_j - d_f \geq r_j + p_j - r_f - p_f$, so $C_j(\pi) - C_f(\pi) < p_j + r_j - p_f - r_f \leq d_j - d_f$. Then, $L_j(\pi,t) < L_f(\pi,t)$ for each job j, $(j \to f)_\pi$.

The inequality (13) can be proved in a similar way.

For each job j satisfying the condition $(j \to s)_\pi$, we have $C_j(\pi) < C_s(\pi)$. If $d_j \geq d_s$, then $L_j(\pi,t) = C_j(\pi) - d_j < C_s(\pi) - d_s = L_s(\pi,t)$, therefore (13) is true.

Let for the job $j \in N \backslash \{f\}$, $(j \to s)_\pi$, $d_j < d_s$, then $r_j > r_s$. Indeed, if we assume that $r_j \leq r_s$, then $r_j(t) \leq r_s(t)$ (it follows from (7)). In addition, $r_s(t) \geq r(t)$ for any job s according to definitions (8) and (11). If $r_s(t) = r(t)$, then for the jobs j and s we can write $r_j(t) = r_s(t) = r(t)$ and $d_j < d_s$, which contradicts the definition (11) of job $s(N,t)$. If $r_s(t) > r(t)$, that is, $r_s > r(t)$, then there is no job $i \in N \backslash \{f, s\}$, for which $r_s > r_i > r(t)$. Therefore, for the jobs j and s we get $r_j(t) = r_s(t)$ and $d_j < d_s$, which contradicts the definition (11) of job $s(N,t)$. Therefore, $r_j > r_s$.

Since $C_j(\pi) \leq C_s(\pi) - p_s$ and $p_j > 0$, then $C_j(\pi) - p_j < C_s(\pi) - p_s$ and since $r_j > r_s$, therefore $C_j(\pi) - p_j - r_j < C_s(\pi) - p_s - r_s$ and

$$C_j(\pi) - C_s(\pi) < p_j + r_j - p_s - r_s. \tag{17}$$

Since $d_j < d_s$, then from (9) we have $d_j - r_j - p_j \geq d_s - r_s - p_s$ or

$$C_j(\pi) - C_s(\pi) < p_j + r_j - p_s - r_s \leq d_j - d_s. \tag{18}$$

Hence, $L_j(\pi) < L_s(\pi)$ for each job $j \in N \backslash \{f\}$, $(j \to s)_\pi$.

Theorem 1.1 If conditions (9) are true for jobs in the subset $N' \subseteq N$, then at any time $t' \geq t$ and any early schedule $\pi \in \Pi(N')$ there exists $\pi' \in \Pi(N')$ such that

$$L_{\max}(\pi', t') \leq L_{\max}(\pi, t'), \quad \text{and} \quad C_{\max}(\pi', t') \leq C_{\max}(\pi, t') \tag{19}$$

and one of the jobs $f = f(N', t')$ or $s = s(N', t')$ is at the first position in schedule π'. If $d_f \leq d_s$, then job f is at the first position in schedule π'.

Proof: Let $\pi = (\pi_1, f, \pi_2, s, \pi_3)$, where π_1, π_2 and π_3 are partial schedules of π. Then, we construct a schedule $\pi' = (f, \pi_1, \pi_2, s, \pi_3)$. From the definitions (7), (8), (10) we get $r_f(t') \leq r_j(t')$, $j \in N'$, hence $C_{\max}((f, \pi_1), t') \leq C_{\max}((\pi_1, f), t')$ and

$$C_{\max}(\pi', t') \leq C_{\max}(\pi, t'), \quad \text{and} \tag{20}$$

$$L_j(\pi', t') \leq L_j(\pi, t'), \quad \forall j \in \{(\pi_2, s, \pi_3)\}. \tag{21}$$

From the lemma 1.1 we have

$$L_j(\pi', t') < L_s(\pi', t'), \quad \forall j \in \{\pi_1\} \cup \{\pi_2\}. \tag{22}$$

Obviously, the following inequality is true for job f

$$L_f(\pi',t') \le L_f(\pi,t'). \tag{23}$$

From (20)–(23) we get $C_{\max}(\pi',t') \le C_{\max}(\pi,t')$ and $L_{\max}(\pi',t') \le L_{\max}(\pi,t')$.

Let $\pi = (\pi_1, s, \pi_2, f, \pi_3)$, that is, job s is before job f. Construct a schedule $\pi' = (s, \pi_1, \pi_2, f, \pi_3)$. Further proof may be repeated as for job f. The first part of the theorem is proved.

Let us assume $d_f \le d_s$ and the schedule $\pi = (\pi_1, s, \pi_2, f, \pi_3)$. Then, we construct a schedule $\pi' = (f, \pi_{11}, \pi_{12}, \pi_3)$, where π_{11}, π_{12} are schedules of jobs of the sets $\{ j \in N' : j \in \{(\pi_1, s, \pi_2)\}, d_j < d_f\}$ and $\{ j \in N' : j \in \{(\pi_1, s, \pi_2)\}, d_j \ge d_f\}$. Jobs in π_{11} and π_{12} are ordered according to nondecreasing release times r_j. From $d_s \ge d_f$ we can conclude that $s \in \{\pi_{12}\}$.

For each job $j \in \{\pi_{11}\}$ we have $d_j < d_f$. Of (9) we get $d_j - r_j - p_j \ge d_f - r_f - p_f$, hence $r_j + p_j < r_f + p_f$, $\forall j \in \{\pi_{11}\}$, and $C_{\max}((f, \pi_{11}), t') = r_f(t') + p_f + \sum_{j \in \{\pi_{11}\}} p_j$. Since jobs in schedule $\{\pi_{12}\}$ are sorted by nondecreasing release times, then $C_{\max}((f, \pi_{11}, \pi_{12}), t') \le C_{\max}((\pi_1, s, \pi_2, f), t')$. As a result

$$C_{\max}(\pi',t') \le C_{\max}(\pi,t'), \quad \text{and} \tag{24}$$

$$L_j(\pi',t') \le L_j(\pi,t'), \quad \forall j \in \{\pi_3\}. \tag{25}$$

Job $j \in \{\pi_{12}\}$ satisfies $d_j \ge d_f$ and $C_j(\pi',t') \le C_f(\pi,t')$, which means

$$L_j(\pi',t') \le L_f(\pi,t'), \quad \forall j \in \{\pi_{12}\}. \tag{26}$$

Since $s \in \{\pi_{12}\}$, then

$$L_s(\pi',t') \le L_f(\pi,t'). \tag{27}$$

From the lemma 1.1

$$L_j(\pi',t') \le L_s(\pi',t'), \quad \forall j \in \{\pi_{11}\}. \tag{28}$$

Moreover, it is obvious that

$$L_f(\pi',t') \le L_f(\pi,t'). \tag{29}$$

From inequalities (24)–(29) it follows that $C_{\max}(\pi',t') \le C_{\max}(\pi,t')$ and $L_{\max}(\pi',t') \le L_{\max}(\pi,t')$, the theorem is proved.

We call a schedule $\pi' \in \Pi(N)$ **effective** if there is no schedule $\pi \in \Pi(N)$ such that $L_{\max}(\pi) \le L_{\max}(\pi')$ and $C_{\max}(\pi) \le C_{\max}(\pi')$, that is, at least one inequality would be strict.

Thus, when constraints (9) are satisfied for jobs from the set N, then there is an effective schedule π', in which either the job $f = f(N,t)$, or $s = s(N,t)$ is present. Moreover, if $d_f \le d_s$, then there is an effective schedule π' with a priority of job f.

We define the set of schedules $\Omega(N,t)$ as a subset of $\Pi(N)$ consisting of $n!$ schedules. Schedule $\pi = (i_1, i_2, \ldots, i_n)$ belongs to $\Omega(N,t)$ if we choose job $i_k, k = 1, 2, \ldots, n$ as $f_k = f(N_{k-1}, C_{i_{k-1}})$ or $s_k = s(N_{k-1}, C_{i_{k-1}})$, where $N_{k-1} = N \backslash \{i_1, i_2, \ldots, i_{k-1}\}$, $C_{i_{k-1}} = C_{i_{k-1}}(\pi)$ and $N_0 = N$, $C_{i_0} = t$. For $d_{f_k} \le d_{s_k}$ it is true that $i_k = f_k$, so if $d_{f_k} > d_{s_k}$, then $i_k = f_k$ or $i_k = s_k$. Its obvious that the set of schedules $\Omega(N,t)$ contains at most 2^n schedules.that is, $p_{2i} > y \ge p_{2i-1}$.

Example 1.1

$$\begin{cases} n = 2m, t \le r_1 < r_2 < \ldots < r_n, \\ r_{2i-1} < r_{2i} + p_{2i} < r_{2i-1} + p_{2i-1}, 1 \le i \le m, \\ r_{2i-1} + p_{2i-1} + p_{2i} < r_{2i+1} < r_{2i} + p_{2i} + p_{2i-1} < r_{2i+2}, 1 \le i \le m-1, \\ r_{2i-1} + p_{2i-1} + p_{2i} - d_{2i-1} > y, 1 \le i \le m-1, \\ r_{2i} + p_{2i} + p_{2i-1} - d_{2i} \le y. \end{cases}$$

The set $\Omega(N,t)$ contains 2^m schedules. The value of y is used below in the text. The optimal solution of the problem $1|r_j, d_j = r_j + p_j, L_{max} \le y|C_{\max}$ is $\pi^* = (2, 1, 4, 3, \ldots, n, n-1)$.

Theorem 1.2 If for the jobs of the subset $N' \subseteq N$, $|N'| = n'$, is true (9), then at any time $t' \ge t$ and any schedule $\pi \in \Pi(N')$ exists a schedule $\pi' \in \Omega(N', t')$ such that

$$L_{\max}(\pi', t') \le L_{\max}(\pi, t') \quad \text{and} \quad C_{\max}(\pi', t') \le C_{\max}(\pi, t'). \tag{30}$$

Proof: Let $\pi = (j_1, j_2, \ldots, j_{n'})$ be an arbitrary schedule. We denote the first l jobs of the schedule π as π_l, $l = 0, 1, 2, \ldots, n'$, where π_0 is an empty schedule, and $\overline{\pi}_l = (j_{l+1}, \ldots, j_{n'})$, then $\pi = (\pi_l, \overline{\pi}_l)$. We introduce $N_l = N' \backslash \{\pi_l\}$ and $C_l = C_{\max}(\pi_l, t')$. Suppose for some l, $0 \le l < n'$, π_l is the largest initial partial the schedule of some schedule from $\Omega(N', t')$. If $j_1 \ne f(N', t')$ and $j_1 \ne s(N', t')$, then $\pi_l = \pi_0$, $l = 0$, then the largest partial schedule is empty. Let us say $f = f(N_l, C_l)$ and $s = s(N_l, C_l)$. If $d_f > d_s$, then $j_{l+1} \ne f$ and $j_{l+1} \ne s$, moreover when $d_f \le d_s$, then $j_{l+1} \ne f$, since π_{l+1} is not an initial schedule of some schedule from $\Omega(N', t')$.

According to the theorem 1.1 for the jobs of the set $\{\overline{\pi}_l\}$, $\overline{\pi}_l \in \Pi(N_l)$, there is a schedule $\overline{\pi}_l'$ starting at time C_l, for which $L_{\max}(\overline{\pi}_l', C_l) \le L_{\max}(\overline{\pi}_l, C_l)$, $C_{\max}(\overline{\pi}_l', C_l) \le C_{\max}(\overline{\pi}_l, C_l)$, and $[\overline{\pi}_l']_1 = (f \text{ or } s)$, moreover, with $d_f \le d_s$, true $[\overline{\pi}_l']_1 = f$, where $[\sigma]_k$ is the job in the k-th place in schedule σ. Hence, $L_{\max}((\pi_l, \overline{\pi}_l'), t') \le L_{\max}((\pi_l, \overline{\pi}_l), t')$ and $C_{\max}((\pi_l, \overline{\pi}_l'), t') \le C_{\max}((\pi_l, \overline{\pi}_l), t')$.

Let us denote $\pi' = (\pi_l, \overline{\pi}_l')$. A feature of schedule π' is that the first $l+1$ jobs are the same as first $l+1$ jobs of some schedule from the set $\Omega(N', t')$, and $L_{\max}(\pi', t') \le L_{\max}(\pi, t')$, $C_{\max}(\pi', t') \le C_{\max}(\pi, t')$.

After no more than n' sequential conversions (since schedule length $n' \le n$) of the original randomly selected schedule π we come to schedule $\pi' \in \Omega(N', t')$, for which $L_{\max}(\pi', t') \le L_{\max}(\pi, t')$ and $C_{\max}(\pi', t') \le C_{\max}(\pi, t')$. The theorem is proved.

We form the following partial schedule $\omega(N,t) = (i_1, i_2, \ldots, i_l)$. For each job i_k, $k = 1, 2, \ldots, l$, we have $i_k = f_k$ and $d_{f_k} \le d_{s_k}$, where $f_k = f(N_{k-1}, C_{k-1})$ and $s_k = s(N_{k-1}, C_{k-1})$. For $f = f(N_l, C_l)$ and $s = s(N_l, C_l)$ inequality $d_f > d_s$ is true. In case when $d_f > d_s$ for $f = f(N,t)$ and $s = s(N,t)$, then $\omega(N,t) = \emptyset$. So $\omega(N,t)$ is the "maximum" schedule, during the construction of which job (like f) for the next place of the schedule can be uniquely selected. We can construct a schedule $\omega(N,t)$ for set of jobs N starting at time t using the algorithm 1.1.

Algorithm 1.1 for constructing schedule $\omega(N,t)$.

1: **Initial step.** Let $\omega = \emptyset$.
2: **Main step.** Find the jobs $f := f(N,t)$ and $s := s(N,t)$;
3: **if** $d_f \le d_s$ **then**
4: $\quad \omega := (\omega, f)$
5: **else**

6: algorithm stops;
7: **end if**
8: Let $N := N \setminus \{f\}, t := r_f(t) + p_f$ and go to the next main step.

Lemma 1.2 The complexity of the algorithm 1.1 for finding the schedule $\omega(N,t)$ is at most $O(n \log n)$ operations for any N and any t.

Proof: At each iteration of the algorithm 1.1 we find two jobs: $f = f(N,t)$ and $s = s(N,t)$. If jobs are ordered by release times r_j (and, accordingly, time $r(t)$ is for $O(1)$ operations), then to find two jobs (f and s) you need $O(\log n)$ operations. Total number of iterations is not more than n. Thus, constructing a schedule $\omega(N,t)$ requires $O(n \log n)$ operations.

The main step of algorithm 1.1 is finding the jobs f and s and it requires at least $O(\log n)$ operations. Obviously, the number of iterations of the algorithm is $O(n)$, therefore, the complexity of the algorithm 1.1 of $O(n \log n)$ operations is the minimum possible for constructing the schedule $\omega(N,t)$.

Lemma 1.3 If for jobs of the set N conditions (9) are true, then any schedule $\pi \in \Omega(N,t)$ starts with the schedule $\omega(N,t)$.

Proof: If $\omega(N,t) = \emptyset$, that is, $d_f > d_s$, where $f = f(N,t), s = s(N,t)$, the statement of the lemma is true, since any schedule starts from empty.

Let $\omega(N,t) = (i_1, i_2, ..., i_l), l > 0$, and so for each $i_k, k = 1, 2, ..., l$, we have $i_k = f_k$ and $d_{f_k} \le d_{s_k}$, where $f_k = f(N_{k-1}, C_{k-1})$ and $s_k = s(N_{k-1}, C_{k-1})$. For $f = f(N_l, C_l)$ and $s = s(N_l, C_l)$ it is true that $d_f > d_s$. As seen from the definition of the set of schedules $\Omega(N,t)$ all schedules in this subset start with a partial schedule $\omega(N,t)$.

Let us use the following notation $\omega^1(N,t) = (f, \omega(N', t'))$ and $\omega^2(N,t) = (s, \omega(N'', t''))$, where $f = f(N,t), s = s(N,t), N' = N \setminus \{f\}, N'' = N \setminus \{s\}, t' = r_f(t) + p_f, t'' = r_s(t) + p_s$. Obviously, the algorithm for finding ω^1 (as well as ω^2) requires $O(n \log n)$ operations, as much as the algorithm for constructing $\omega(N,t)$.

Consequence 1.1 **from Lemma 1.3.** If the jobs of the set N satisfy conditions (9), then each schedule $\pi \in \Omega(N,t)$ starts either with $\omega^1(N,t)$, or with $\omega^2(N,t)$.

Theorem 1.3 If the jobs of the set N satisfy conditions (9), then for any schedule $\pi \in \Omega(N,t)$ it is true that $(i \to j)_\pi$ for any $i \in \{\omega^1(N,t)\}$ and $j \in N \setminus \{\omega^1(N,t)\}$.

Proof: In the case $\{\omega^1(N,t)\} = N$ statement of the theorem is obviously true. Let $\{\omega^1(N,t)\} \ne N$. Further in in the proof we use the notation $\omega^1 = \omega^1(N,t)$.

If $f = f(N,t)$ and $s = s(N,t)$ are such that $d_f \le d_s$, then all schedules from the set $\Omega(N,t)$ begin with a partial schedule $\omega(N,t) = \omega^1$, therefore the statement of the theorem is also true.

Consider the case of $d_f > d_s$. All schedules from set $\Omega(N,t)$ starting with job f have partial schedule $\omega(N,t) = \omega^1$.

Let us choose any arbitrary schedules $\pi \in \Omega(N,t)$ with job s comes first, $\pi_1 = s$, and any schedule $|\omega^1| = l, l < n$, containing l jobs. Let $\pi_l = (j_1, j_2, ..., j_l)$ be a partial schedule of schedule π containing l jobs, where $j_1 = s$. We need to prove that $\{\pi_l\} = \{\omega^1\}$. Let us assume the contrary that there is a job $j \in \{\pi_l\}$, but $j \notin \{\omega^1\}$.

For case $(j \to f)_\pi$ we need to check two subcases. If $d_j < d_f$, then from (9) we have $d_j - r_j - p_j \ge d_f - r_f - p_f$, therefore $r_j + p_j < r_f + p_f$. Then job j is included in schedule ω^1 according to the definition of $\omega(N,t)$ and ω^1, but by our assumption $j \notin \{\omega^1\}$. If $d_j \ge d_f$, then from the fact that $\pi \in \Omega(N,t)$ follows $(f \to j)_\pi$, but this contradicts $(j \to f)_\pi$. Therefore, $j \in \{\omega^1\}$.

The other case is $(f \to j)_\pi$. Then for each job $i \in \{\omega^1\}$, for which $i \notin \{\pi_l\}$, conditions $r_i < r_i + p_i \le C_{\max}(\omega^1) < r_{s_{l+1}} \le r_j$ are true, because $j \notin \{\omega^1\}$, where $s_{l+1} = s(N \backslash \{\omega^1\}, C_{\max}(\omega^1))$. Jobs s_{l+1} and j were not ordered in schedule ω^1, therefore, $C_{\max}(\omega^1) < r_{s_{l+1}} \le r_j$. Besides, $d_i \le d_j$. If $d_i > d_j$, then $r_i + p_i \ge r_j + p_j$, but $r_i + p_i < r_j$ is true. Hence $(i \to j)_{\pi_l}$, since $\pi = (\pi_l, \overline{\pi}_l) \in \Omega(N, t)$, but it contradicts our guess that $i \notin \{\pi_l\}$ and $j \in \{\pi_l\}$.

Therefore, our assumption is not true and $\{\omega^1\} = \{\pi_l\}$. The theorem is proved.

Therefore, jobs of the set $\{\omega^1(N,t)\}$ precede jobs of the set $N \backslash \{\omega^1(N,t)\}$ for any schedule from the set $\Omega(N,t)$, including the optimal schedule.

3.2 Performance problem with constraint on maximum lateness

The problem $1|d_i \le d_j, d_i - r_i - p_i \ge d_j - r_j - p_j; L_{\max} \le y|C_{\max}$ consists of the following. We need to find schedule θ for any y with $C_{\max}(\theta) = \min\{C_{\max}(\pi) : L_{\max}(\pi) \le y\}$. If $L_{\max}(\pi) > y$ for any $\pi \in \Pi(N)$, then $\theta = \emptyset$.

Lemma 1.4 The complexity of algorithm 1.2 does not exceed $O(n^2 \log n)$ operations.

Proof: At each iteration of the main step of the algorithm 1.2 we find the schedules ω^1 and, if necessary, ω^2 in $O(n \log n)$ operations. Since ω^1 and ω^2 consist of at least one job, then at each iteration of the algorithm we either add one or mere jobs to the schedule θ, or assume $\theta = \emptyset$ and stop. Therefore, the total number of steps of the algorithm is at most n. Thus, algorithm 1.2 requires $O(n^2 \log n)$ operations.

Algorithm 1.2 for solving the problem $1|d_i \le d_j, d_i - r_i - p_i \ge d_j - r_j - p_j; L_{\max} \le y|C_{\max}$.

Initial step. Let $\theta := \omega(N,t), t' := t$;
Main step.
if $L_{\max}(\theta, t') > y$ **then**
  $\theta := \emptyset$ and the algorithm stops.
end if
Find $N' := N \backslash \{\theta\}, t' := C_{\max}(\theta)$ and $\omega^1(N', t'), \omega^2(N', t')$.
if $N' = \emptyset$ **then**
  the algorithm stops.
else
  if $L_{\max}(\omega^1, t') \le y$ **then**
    $\theta := (\theta, \omega^1)$ and go to next step;
  end if
  if $L_{\max}(\omega^1, t') > y$ and $L_{\max}(\omega^2, t') \le y$ **then**
    $\theta := (\theta, \omega^2)$ and go to next step;
  end if
  if $L_{\max}(\omega^1, t') > y$ and $L_{\max}(\omega^2, t') > y$ **then**
    $\theta := \emptyset$ and the algorithm stops.
  end if
end if

The problem $1|d_i \le d_j, d_i - r_i - p_i \ge d_j - r_j - p_j; L_{\max} \le y|C_{\max}$ cannot be solved in less than $O(n^2 \log n)$ operations because there exists (*Example 1.1*). The optimal schedule for this example is $\pi^* = (2, 1, 4, 3, \ldots, n, n-1)$. To find this schedule, we need $O(n^2 \log n)$ operations.

We denote by $\theta(N,t,y)$ the schedule constructed by algorithm 1.2 starting at time t from the jobs of the set N with the maximum lateness not more than y. If $N = \emptyset$, then $\theta(\emptyset, \text{t}, \text{y}) = \emptyset$ for any t and y.

Theorem 1.4 Let the jobs of the set N satisfy conditions (9). If the algorithm 1.2 constructs the schedule $\theta(N,t,y) \neq \emptyset$, then $C_{\max}(\theta) = \min\{C_{\max}(\pi) : L_{\max}(\pi) \leq y, \pi \in \Pi(N)\}$. If, as a result of the algorithm 1.2 the schedule will not be generated, that is, $\theta(N,t,y) = \emptyset$, then $L_{\max}(\pi) > y$ for each $\pi \in \Pi(N)$.

Proof: In case if for schedule $\pi \in \Pi(N)$ condition $L_{\max}(\pi) \leq y$ is true, then according to Theorem 1.2 there is a schedule $\pi' \in \Omega(N,t)$ such that $L_{\max}(\pi') \leq L_{\max}(\pi) \leq y$ and $C_{\max}(\pi') \leq C_{\max}(\pi)$. Therefore, the required schedule θ contains in set $\Omega(N,t)$.

According to Lemma 1.3, all schedules of the set $\Omega(N,t)$ start with $\omega(N,t)$. Let us take $\theta_0 = \omega(N,t)$.

After $k, k \geq 0$ main steps of the algorithm 1.2 we got the schedule θ_k and $N' = N \backslash \{\theta_k\}, t' = C_{\max}(\theta_k)$. Let us assume that there is an optimal by the criterion of maximum completion time ($C_{\max}$) schedule θ starting with θ_k. According to Theorem 1.2, there is an optimal extension of the schedule θ_k among the schedules from the set $\Omega(N', t')$.

Let $\theta_{k+1} = (\theta_k, \omega^1(N', t'))$, that is, $L_{\max}(\theta_{k+1}) \leq y$. According to Theorem 1.3, for schedule ω^1, $\omega^1 = \omega^1(N', t')$, there is no artificial idle times of the machine and all schedules from the set $\Omega(N', t')$ start with jobs of the set $\{\omega^1(N', t')\}$. Therefore, $\omega^1(N', t')$ is the best by the criterion of $C_{\max}$ among all feasible by maximum lateness ($L_{\max}$) extensions of the partial schedule θ_k.

If $\theta_{k+1} = (\theta_k, \omega^2(N', t'))$, that is, $L_{\max}(\omega^1, t') > y$, and $L_{\max}(\omega^2, t') \leq y$. All schedules of the set $\Omega(N', t')$ start with either schedule $\omega^1(N', t')$ or $\omega^2(N', t')$. As $L_{\max}(\omega^1, t') > y$, then the only suitable extension is $\omega^2(N', t')$.

Thus, at each main step of the algorithm, we choose the fastest continuation of the partial schedule θ_k among all those allowed by the maximum lateness. After no more than n main steps of the algorithm, the required schedule is constructed.

Let us assume that after the $k+1$ steps of the algorithm $L_{\max}(\omega^1, t') > y$ and $L_{\max}(\omega^2, t') > y$. If schedule θ could exist, that is, $\theta \neq \emptyset$, then θ would start with θ_k. Then for any schedule $\pi \in \Pi(N', t')$ there would exist a schedule $\pi' \in \Omega(N', t')$ such that $L_{\max}(\pi, t') \geq L_{\max}(\pi', t') \geq L_{\max}(\omega^1, t') > y$ or $L_{\max}(\pi, t') \geq L_{\max}(\pi', t') \geq L_{\max}(\omega^2, t') > y$. Therefore $\theta = \emptyset$.

Repeating our proof as many times as the main step of algorithm 1.2 (no more than n), we come to the truth of the statement of the theorem.

3.3 Algorithm for constructing a set of Pareto schedules by criteria $C_{\max}$ and $L_{\max}$

Let us develop an algorithm for constructing a set of Pareto schedules $\Phi(N,t) = \{\pi'_1, \pi'_2, \ldots, \pi'_m\}$, $m \leq n$, by criteria $C_{\max}$ and $L_{\max}$ according to conditions (5)–(6).

Schedule π'_m is a solution to problem $1|r_j|L_{\max}$ if (9) is true.

Algorithm 1.3 for constructing a set of Pareto schedules by criteria $C_{\max}$ and $L_{\max}$.

1: **Initial step.** $Y := +\infty$, $\pi^* := \omega(N,t)$, $\Phi := \emptyset$, $m := 0$, $N' := N \backslash \{\pi^*\}$ and $t' := C_{\max}(\pi^*)$.

2: **if** $N' = \emptyset$ **then**

3: $\Phi := \Phi \cup (\pi^*)$, $m := 1$ and the algorithm stops.

4: **end if**

5: **Main step.**

if $L_{\max}(\omega^1, t') \le L_{\max}(\pi^*)$ **then**
  $\pi^* := (\pi^*, \omega^1)$, where $\omega^1 = \omega^1(N', t')$ and go to the next step;
end if
if $L_{\max}(\omega^1, t') > L_{\max}(\pi^*)$ **then**
  if $L_{\max}(\omega^1, t') < y$ **then**
    find $\theta = \theta(N', t', y')$ using algorithm 1.2, where $y' = L_{\max}(\omega^1, t')$;
    if $\theta = \emptyset$ **then**
      $\pi^* := (\pi^*, \omega^1)$ and go to the next step;
    else
      $\pi' := (\pi^*, \theta)$
      if $C_{\max}(\pi'_m) < C_{\max}(\pi')$ **then**
        $m := m + 1$, $\pi'_m := \pi'$, $\Phi := \Phi \cup (\pi'_m)$, $y = L_{\max}(\pi'_m)$;
      else
        $\pi'_m = \pi'$ and go to next step;
      end if
    end if
    if $L_{\max}(\omega^1, t') \ge y$ **then**
      find $\omega^2 = \omega^2(N', t')$;
      if $L_{\max}(\omega^2, t') < y$ **then**
        $\pi^* = (\pi^*, \omega^2)$ and go to the next step;
      else
        $\pi^* = \pi_m{}'$ and the algorithm stops.
      end if
    end if
  end if
end if

As a result of the algorithm 1.3, a set of schedules $\Phi(N, t)$ is constructed, for the set of jobs N starting at time t, for which inequality $1 \le |\Phi(N, t)| \le n$ true. We should note that the set $\Phi(N, t)$ for *Example 1.1* consists of two schedules, although set $\Omega(N, t)$ consists of $2^{\frac{n}{2}}$ schedules:

$$\pi_{1'} = (1, 2, 3, 4, \ldots, n-1, n), \tag{31}$$

$$\pi_{2'} = (2, 1, 4, 3, \ldots, n, n-1). \tag{32}$$

Lemma 1.5 The complexity of the algorithm 1.3 does not exceed $O(n^3 \log n)$ operations.

Proof: At each iteration of the main step of the algorithm 1.3 we find schedules ω^1 and, if necessary, ω^2, which requires $O(n \log n)$ operations according to lemma 1.2, and also schedule θ in $O(n^2 \log n)$ operations. As ω^1 and ω^2 consist of at least one job, then at any iteration of the algorithm one or more jobs are added to the schedule π^*, or the algorithm stops at last schedule π'. Therefore, the total number of iterations is at most n. Thus, it takes no more than $O(n^3 \log n)$ operations to execute algorithm 1.3.

Theorem 1.5 If case if (9) is true for each job of the set N, then the schedule π^*, constructed by algorithm 1.3, is optimal according to the criterion $L_{\max}$. Moreover, for any schedule $\pi \in \Pi(N)$ there exists a schedule $\pi' \in \Phi(N, t)$ such that $L_{\max}(\pi') \le L_{\max}(\pi)$ and $C_{\max}(\pi') \le C_{\max}(\pi)$.

Proof: According to Theorem 1.2, there exists an optimal (by $L_{\max}$) schedule from set $\Omega(N, t)$. According to Lemma 1.3, all schedules of the set $\Omega(N, t)$ start with a partial schedule $\omega(N, t)$.

Let $\pi_0 = \omega(N,t)$. After k, $k \geq 0$, main steps of algorithm 1.3 we have a partial schedule π_k. Suppose there is an optimal (by $L_{\max}$) schedule starting with π_k. We denote $N' = N \backslash \{\pi_k\}$ and $t' = C_{\max}(\pi_k)$.

If $\pi_{k+1} = (\pi_k, \omega^1)$, where $\omega^1 = \omega^1(N't')$, then either $L_{\max}(\omega^1, t') \leq L_{\max}(\pi_k)$, or $L_{\max}(\pi_k) < L_{\max}(\omega^1, t') < y$, that is, current value of the criterion and the maximum lateness will "appear" on next steps of the algorithm 1.3. That is, $\theta(N', t', y') = \emptyset$, where $y' = L_{\max}(\omega^1, t')$. If $\theta = \theta(N', t', y') \neq \emptyset$, then we improve the current maximum lateness value: $\pi' = (\pi_k, \theta)$ and $y = L_{\max}(\pi') = L_{\max}(\omega^1, t')$. The schedule π' is added to the set of schedules $\Phi(N,t)$. Moreover, according to Theorem 1.3 jobs of set $\{\omega^1\}$ precede jobs of set $N' \backslash \{\omega^1\}$. Thus, the schedule ω^1alert(without artificial idle times of the machine) would be the best continuation for π_k.

If $\pi_{k+1} = (\pi_k, \omega^2)$, where $\omega^2 = \omega^2(N', t')$, that is, according to algorithm 1.3 $L_{\max}(\omega^2, t') < L_{\max}(\pi') \leq L_{\max}(\omega^1, t')$. In this case the continuation ω^2 is "better" than ω^1. Hence, the partial schedule π_{k+1} is a part of some optimal schedule.

Repeating our proof no more than n times, we come to optimality (for $L_{\max}$) of the schedule π^*.

The set of schedules $\Phi(N,t)$ contains at most n schedules, since at each main step of the algorithm in the set $\Phi(N,t)$ at most one schedule is "added," and this step is executed no more than n times.

Suppose there is a schedule $\pi \in \Pi(N), \pi \notin \Phi(N,t)$, such that either $C_{\max}(\pi) \leq C_{\max}(\pi')$ and $L_{\max}(\pi) \geq L_{\max}(\pi')$, or $C_{\max}(\pi) \geq C_{\max}(\pi')$ and $L_{\max}(\pi) \leq L_{\max}(\pi')$ for each schedule $\pi' \in \Phi(N,t)$. Moreover, in each pair of inequalities at least one inequality is strict. According to Theorem 1.1, there is a schedule $\pi'' \in \Omega(N,t)$ such that $L_{\max}(\pi'') \leq L_{\max}(\pi)$ and $C_{\max}(\pi'') \leq C_{\max}(\pi)$. If $\pi'' \in \Phi(N,t)$. Thus, it becomes obvious that our assumption is not correct. Let $\pi'' \in \Omega(N,t) \backslash \Phi(N,t)$. Algorithm 1.3 shows that the structure of each schedule $\pi' \in \Phi(N,t)$ can be represented as a sequence of partial schedules $\pi' = \left(\omega'_0, \omega'_1, \omega'_2, \ldots, \omega'_{k'}\right)$, where $\omega'_0 = \omega(N,t)$, and ω'_i is either $\omega^1(N'_i, C'_i)$, or $\omega^2(N'_i, C'_i)$, and $N'_i = N \backslash \{\omega'_0, \ldots, \omega'_{i-1}\}$, $C'_i = C_{\max}((\omega'_0, \ldots, \omega'_{i-1}), t), i = 1, 2, \ldots, k'$. The schedule π'' has the same structure according to the definition of the set $\Omega(N,t)$, that is, $\pi = (\omega''_0, \omega''_1, \omega''_2, \ldots, \omega''_{k''})$, possibly $k'' \neq k'$, where $\omega''_0 = \omega'_0 = \omega(N,t), \omega''_i$ is equal to either $\omega^1(N''_i, C''_i)$, or $\omega^2(N''_i, C''_i)$, a $N''_i = N \backslash \{\omega''_0, \ldots, \omega''_{i-1}\}$, $C''_i = C_{\max}((\omega''_0, \ldots, \omega''_{i-1}), t)$, $i = 1, 2, \ldots, k''$.

We assume that the first k partial schedules π'' and π' are equal, that is, $\omega''_i = \omega'_i = \omega_i, i = 0, 1, \ldots, k-1, \omega''_k \neq \omega'_k$. If $y = L_{\max}(\omega_0, \ldots, \omega_{k-1})$, let us construct a schedule θ using algorithm 1.2, $\theta = \theta(N_k, C_k, y)$. If $\theta = \emptyset$, then according to algorithm 1.3, $\omega'_k = \omega^1(N_k, C_k)$. Because of $\omega''_k \neq \omega'_k$, schedule $\omega''_k = \omega^2(N_k, C_k)$. Objective function value ($L_{\max}$) can be reached on a job from the set N_k, since $\theta = \emptyset$. The whole structure of the algorithm 1.3 construct in such a way that up to the "critical" job (according to $L_{\max}$) order the jobs as "tightly" as possible, therefore we complete the schedule ω^1, after which $C_{\max}(\pi') \leq C_{\max}(\pi'')$ and $L_{\max}(\pi') \leq L_{\max}(\pi'')$. If $\theta \neq \emptyset$, then for schedules π' and π'' $C_{\max}(\pi') \leq C_{\max}(\pi'')$ and $L_{\max}(\pi') = L_{\max}(\pi'')$. Thus, for any schedule $\pi'' \in \Omega(N,t) \backslash \Phi(N,t)$ exists schedule $\pi' \in \Phi(N,t)$ such that $C_{\max}(\pi') \leq C_{\max}(\pi'')$ and $L_{\max}(\pi') \leq L_{\max}(\pi'')$. Hence, for any schedule $\pi \in \Pi(N)$ there exists schedule $\pi' \in \Phi(N,t)$ such that $L_{\max}(\pi') \leq L_{\max}(\pi)$ and $C_{\max}(\pi') \leq C_{\max}(\pi)$. The theorem is proved.

Figure 1 schematically shows the considered schedule.

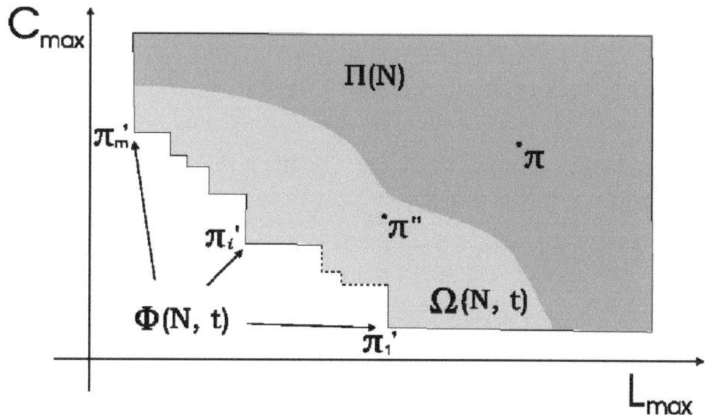

Figure 1.
The set of Pareto-optimal schedules.

For the set of schedules $\Phi(N,t) = \{\pi'_1, \pi'_2, \ldots, \pi'_m\}$, $m \leq n$, we conditions (5)–(6) are true.

The schedule π'_1 is optimal in terms of speed ($C_{\max}$), and π'_m is optimal in terms of the maximum lateness (by $L_{\max}$) if the jobs of the set N satisfy the conditions (9).

4. Conclusions

Single machine scheduling problem with given release dates and two objective functions is considered in this chapter, which is *NP*-hard in the strong sense. A number of new polynomially and pseudo-polynomially solvable subcases of the problem were found. For a case when

$$d_1 \leq \ldots \leq d_n, \quad d_1 - r_1 - p_1 \geq \ldots \geq d_n - r_n - p_n, \tag{33}$$

an algorithm for constructing a Pareto-optimal set of schedules by criteria $C_{\max}$ and $L_{\max}$ is developed. It is proved that the complexity of the algorithm does not exceed $O(n^3 \log n)$ operations.

An experimental study of the algorithm showed that it can be used to construct optimal schedules (by $L_{\max}$) even for instances not satisfying the conditions (33).

Acknowledgements

The research was supported by RFBR (project 20-58-S52006).

Author details

Alexander A. Lazarev and Nikolay Pravdivets*
Institute of Control Sciences, Moscow, Russia

*Address all correspondence to: pravdivets@ipu.ru

References

[1] Lenstra JK, Rinnooy Kan AHG, Brucker P. Complexity of machine scheduling problems. Annals of Discrete Mathematics. 1977;**1**:343-362

[2] Graham RL, Lawler EL, Lenstra JK, Rinnooy Kan AHG. Optimization and approximation in deterministic sequencing and scheduling: A survey. Annals of Discrete Mathematics. 1979;**5**: 287-326

[3] Potts CN. Analysis of a heuristic for one machine sequencing with release dates and delivery times. Operations Research. 1980;**28**:1436-1441

[4] Jackson JR. Scheduling a Production Line to Minimize Maximum Tardiness. Los Angeles, CA: University of California; 1955. Manag. Sci. Res. Project. Research Report N 43

[5] Hall LA, Shmoys DB. Jackson's rule for one-machine schedulings: Making a good heuristic better. Mathematics of Operations Research. 1992;**17**:22-35

[6] Mastrolilli M. Efficient approximation schemes for scheduling problems with release dates and delivery times. Journal of Scheduling. 2003;**6**(6):521-531

[7] Garey MR, Johnson DS, Simons BB, Tarjan RE. Scheduling UnitTime tasks with arbitrary release times and deadlines. SIAM Journal on Computing. 1981;**10**:256-269

[8] Vakhania N. Single-machine scheduling with release times and tails. Annals of Operations Research. 2004; **129**(1–4):253-271

[9] Vakhania N. Dynamic restructuring framework for Scheduling with release times and due-dates. Mathematics. 2019;**7**(11):1104

[10] Lawler EL. Optimal sequencing of a single machine subject to precedence constraints. Management Science. 1973; **19**(5):544-546

[11] Simons BB. A fast algorithm for single processor scheduling. In: Proceedings of the 19th IEEE Annual Symposium on Foundations of Computer Science. New York: Ann. Arbor. Mich; 1978. pp. 246-252

[12] Baker KR, Lawler EL, Lenstra JK, Rinnooy Kan AHG. Preemtive scheduling of a single machine to minimize maximum cost subject to release dates and precedence constraints. Operations Research. 1983; **31**(2):381-386

[13] Hoogeveen JA. Minimizing maximum promptness and maximum lateness on a single machine. Mathematics of Operations Research. 1996;**21**:100-114

[14] Lazarev AA, Shulgina ON. Polynomially solvable subcases of the problem of minimizing maximum lateness. Izvestiya VUZov. Mathematics. 2000 (in Russian)

[15] Vakhania N. Scheduling a single machine with primary and secondary objectives. Algorithms. 2018;**11**(6):80

[16] Vakhania N. Fast solution of single-machine scheduling problem with embedded jobs. Theoretical Computer Science. 2019;**782**:91-106

Section 2

Non-Traditional Approach: Threshold Optimality

Chapter 3

A Brief Look at Multi-Criteria Problems: Multi-Threshold Optimization versus Pareto-Optimization

Nodari Vakhania and Frank Werner

Abstract

Multi-objective optimization problems are important as they arise in many practical circumstances. In such problems, there is no general notion of optimality, as there are different objective criteria which can be contradictory. In practice, often there is no unique optimality criterion for measuring the solution quality. The latter is rather determined by the value of the solution for each objective criterion. In fact, a practitioner seeks for a solution that has an acceptable value of each of the objective functions and, in practice, there may be different tolerances to the quality of the delivered solution for different objective functions: for some objective criteria, solutions that are far away from an optimal one can be acceptable. Traditional Pareto-optimality approach aims to create all non-dominated feasible solutions in respect to all the optimality criteria. This often requires an inadmissible time. Besides, it is not evident how to choose an appropriate solution from the Pareto-optimal set of feasible solutions, which can be very large. Here we propose a new approach and call it multi-threshold optimization setting that takes into account different requirements for different objective criteria and so is more flexible and can often be solved in a more efficient way.

Keywords: multi-critneria optimization, optimal solution, Pareto-optimization, multi-threshold optimization, scheduling algorithm, time complexity

1. Introduction

Multi-objective optimization problems are important as they arise in many practical circumstances. In such problems, there is no general notion of optimality, as there are different objective criteria which are often contradictory: an optimal solution for one criterion may be far away from an optimal one for some other criterion. Thus for many such real-life problems, there is no unique optimality criterion for measuring the solution quality. The latter is rather determined by the value of the solution for each objective criterion. In fact, a practitioner is not interested, generally, in optimizing a particular objective criterion, but he rather seeks for a solution that has an acceptable value of each of the objective functions. Furthermore, in practice, there may exist different tolerances to the quality of the delivered solution for different objective functions. In particular, for some objective criteria, solutions far away from an optimal one can be acceptable. Such solutions can often be obtained by relatively low computational efforts even for intractable problems.

Taking into account these considerations, here we propose a new approach and call it multi-threshold optimization setting that takes into account different requirements for different objective criteria, in contrary to a traditional Pareto-optimality approach. The Pareto-optimality concept, named after an Italian scientist Vilfredo Pareto, is a traditionally used compromise to address a complicated multi-objective scenario. It looks for a so-called Pareto-optimal frontier of the feasible solutions consisting of those solutions that are not dominated by any other feasible solution (with respect to any of the given objective functions). This often requires an inadmissible time: finding the Pareto-optimal frontier often remains an intractable (NP-hard) problem. This is always the case if at least one of the corresponding single-criterion problems is NP-hard. Finding the Pareto-optimal set of solutions may be NP-hard even if none of the single-criterion problem is NP-hard. Besides, it is not evident how to choose an appropriate solution from the Pareto-optimal set of feasible solutions, which can be very large. The multi-threshold optimization approach is more flexible since it takes into account different requirements for different objective criteria: in practice, some objective criteria can be more critical than the other ones, and hence there may exist different degrees of tolerance for the deviation of the objective value of different criteria from the optimal objective value of the corresponding single-criterion problems.

The multi-threshold optimization problem seeks for a feasible schedule whose objective values are acceptable for a given particular application for all objective functions; in particular, they do not exceed (for minimization problems) or are no smaller (for maximization problems) than the components of a threshold vector specified by the practitioner whose ith component is some threshold value for the ith objective function. As we observe, depending on the components of the above vector, it might be possible to solve the multi-threshold optimization problem in a low-degree polynomial time even if all the corresponding single-criterion problems are NP-hard. A threshold vector with specific threshold values for each objective function is supposed to have a direct practical meaning. For practically useful values of the threshold vector, the multi-threshold optimization problem might be solved in a low-degree polynomial time by a kit of heuristic algorithms, each one being designed for one of the corresponding single-criterion problems. If the kit of heuristic algorithms fails to find a feasible solution respecting the threshold vector, then the heuristics for NP-hard single-criterion problems can be replaced by implicit enumeration algorithms. In fact, the replacement can be accomplished step by step, starting from the most critical heuristics. This kind of approach may be more practical since some objective criteria can be optimized easier than other ones. Besides, as already noted, the practitioner may not be interested, in general, in the minimization of each objective function but rather in a solution of an acceptable quality for every objective function: in practice, there may be different tolerances to the quality of the delivered solution for each objective function, and different objective functions might be optimized with quite different costs.

Thus our approach may lead to more efficient and practical solution of a multi-criteria problem than the corresponding Pareto-optimal setting. In the following sections, we give a brief comparative analysis of the Pareto-optimization approach with our multi-threshold optimization approach illustrating its advantage on single-machine scheduling problems.

2. Multi-criteria optimization problems

For an extensive description of multi-criteria optimization problems and the solution methods, the reader may have a look on a book by T'kindt and Billaut [1] and a survey paper [2] by the same authors.

Discrete *optimization* problems have emerged in the late 1940s of the last century due to the rapid growth of the industry and new rising demands in efficient solution methods. Modeled in mathematical language, such an optimization problem has a finite set of so-called feasible solutions; each feasible solution is determined by a set of mathematically formulated restrictions that naturally arise in practice. The quality of a feasible solution is measured by an objective function, whose domain is the whole set of feasible solutions. Ideally, one aims to determine a feasible solution that gives an extremal (minimal or maximal) value to the objective function, a so-called optimal solution. Since the number of feasible solutions is typically finite, theoretically, finding an optimal solution is trivial: just enumerate all the feasible solutions calculating for each of the value of the objective function and select any one with the optimal objective value. The main issue here is that a complete enumeration of all feasible solutions is mostly impossible in practice.

There are two distinct classes of combinatorial optimization problems, the class P of polynomially solvable ones and the intractable NP-hard problems. For a problem from the class P, there exists an efficient (polynomial in the size of the problem) algorithm, whereas no such algorithm exists for an NP-hard problem (the number of feasible solutions of an NP-hard optimization problem grows exponentially with the size of the input). It is widely believed that it is very unlikely that an NP-hard problem can be solved in polynomial time. Hence, it is natural to develop approximation solution methods.

Multi-criteria optimization problems are optimization problems with two or more different objective criteria. For the majority of such problems, there exists no single solution which optimizes (minimizes or maximizes) all the objective functions. In this sense, different objectives are contradictory, and hence, it is not straightforward to understand which feasible solution to the problem is optimal: a multi-criteria optimization problem typically has no optimal solution. In this situation, one may look for a solution which attains an acceptable value for each objective function or a solution which is not dominated by any other solution, in the sense that there is no other feasible solution which attains better objective values for all objective functions. We shall refer to the first and second versions of the multi-criteria optimization problem as *multi-threshold optimization* and *Pareto-optimization* versions and define them more formally below.

Let the k objective functions over the set $\mathcal{F}$ of feasible solutions of a given multi-criteria optimization problem be $f_1, \dots, f_k$. Since these functions might be mutually contradictory, there may exist no feasible solution minimizing/maximizing all objective functions simultaneously. Without loss of generality, let us consider from now on the minimization version of our multi-criteria optimization problem.

Let F_i^* be the optimal value for a single-criterion problem with the objective to minimize function f_i, and let A^i be some threshold value for the objective function f_i, $i = 1, \dots, k$.

In the multi-threshold optimization version, we look for a feasible solution σ such that $f_i(\sigma) \leq A^i$ for each $i = 1, \dots, k$.

A commonly used dominance relation for the Pareto-optimization version is defined as follows.

A solution $\sigma_1 \in \mathcal{F}$ *dominates* a solution $\sigma_2 \in \mathcal{F}$ if $f_i(\sigma_1) < f_i(\sigma_2)$ for $i = 1, \dots, k$; in fact, we allow $\leq$ instead of $<$ for all values of i except one requiring to have at least one strict inequality.

Now $\sigma \in \mathcal{F}$ is a *Pareto-optimal solution* if no other solution from the set $\mathcal{F}$ dominates the solution σ. We shall refer to the set of all such feasible solutions as *Pareto-optimal set*. Forming a Pareto-optimal set of feasible solutions may be not easy. For instance, let, for $k = 2, f_1(\sigma_1) \leq f_1(\sigma_2)$; then solution σ_2 is dominated by solution σ_1 if $f_2(\sigma_1) < f_2(\sigma_2)$. This condition can be verified in polynomial time for any pair of solutions σ_1 and σ_2 (given that the corresponding optimization problem is from the class NP). However, whenever the number of feasible solutions grows exponentially with the length of the input (which is the case for NP-hard problems), the explicit evaluation of all possible pairs of feasible solutions (which is unavoidable for finding a dominant solution) would lead us to an exponential-time performance. In particular, if one of the single-criterion problems is NP-hard, finding a Pareto-optimal set for the multi-objective setting will take an exponential time.

Theorem 1 The problem of finding a Pareto-optimal set of feasible solutions for a multi-objective optimization problem with the objective functions $f_1, \dots, f_k$ is NP-hard if one of the corresponding single-criterion problems is NP-hard.

Proof. We basically reformulate the above reasoning. Consider a bi-criteria optimization problem with $k = 2$. Consider the set S_A of feasible solutions with $f_1(\sigma) = A$ for all $\sigma \in S_A$ and some threshold value A of function f_1 (without loss of generality assume that $S_A \neq \emptyset$). A Pareto-optimal solution from the set S_A must attain the minimum possible value of function f_2 as otherwise it will be dominated by one that attains this value. Then we arrive at a single-criterion optimization problem with the objective function f_2, which is NP-hard.

From the first glance, the multi-threshold optimization version of a multi-criteria optimization problem may seem to be easier than the Pareto-optimality version. This is, in part, correct, but considering a threshold vector with arbitrary components, in general, we will also arrive at an intractable problem as the decision version of an NP-hard single-criterion optimization problem is NP-complete. In particular, suppose that we are given a single-criterion optimization problem with the objective to minimize the function f_i ($i \in \{1, \dots, k\}$). If this problem is NP-hard, then its decision version, given a threshold value A of function f_i, if there is a feasible solution $\sigma \in \mathcal{F}$ with $f_i(\sigma) \leq A$, is NP-complete. Hence, if one of the single-criterion optimization problems is NP-hard, then the multi-threshold optimization version of the corresponding multi-criteria optimization problem is also NP-hard.

At the same time, finding a Pareto-optimal set of feasible solutions may be NP-hard even if none of the single-criterion problem is NP-hard, i.e., they are solvable in polynomial time. Can the multi-threshold optimization version of a multi-criteria optimization problem be solved in polynomial time, if all the corresponding single-criterion optimization problems are polynomial? In other words, suppose that the single-criterion problem of finding a feasible solution attaining the minimum value of the objective function f_i for $i = 1, \dots, k$ can be solved in polynomial time. Then clearly, the decision version that seeks for a feasible solution $\sigma \in \mathcal{F}$ with $f_i(\sigma) \leq A$ is also polynomially solvable.

Unlike the Pareto-optimization problem, the multi-threshold optimization problem may be solvable in polynomial time even if all the corresponding single-criterion problems are NP-hard; whether it is solvable in polynomial time or not essentially depends on the particular threshold vector $\mathbf{A} = (A^1, \dots, A^k)$. As we shall argue in the next sections, depending on the particular threshold values for each objective function, it might be possible to solve the multi-threshold optimization problem in a low-degree polynomial time even if all the corresponding single-criterion problems are NP-hard. The given threshold values for each objective function may have a direct practical meaning. For practically useful values of the threshold vector **A**, the corresponding instance of the multi-threshold optimization problem might be solved in a low-degree polynomial time though it may be NP-hard, in general (for an arbitrary threshold vector, see Section 3).

3. Some basic single-criterion scheduling problems

In the rest of this chapter, we illustrate the Pareto-optimality and the multi-threshold optimization approaches for *scheduling problems*. For recent developments in multi-criteria optimization for scheduling problems, the reader is referred to a recent survey by Nagar et al. [3] and Parveen and Ullah [4] and for some earlier works approximately until the year 2005 to the earlier cited work by T'kindt and Billaut [1].

The scheduling problems arise in various practical circumstances. Examples of such problems are job shop problems in industry, scheduling of information and computational processes, and traffic scheduling and servicing of cargo trains, ships, and airplanes. There are scheduling problems of diverse types and different complexities. Saying generally, one deals with two primary notions: *job* (or *task*) and *machine* (or *processor*). A job is a part of the whole work to be done; a machine is the means for the performance of a job. A common restriction in scheduling problems is that a machine cannot handle more than one job at a time. Each job j is characterized by its *processing time* p_j, i.e., it needs this prescribed time on a machine. A job may have other parameters as well, which may yield additional restrictions and/or can be employed by an objective function. For instance, the *release time* r_j of job j is the time moment when job j becomes available (it cannot be scheduled before that time). The *due date* d_j of job j is the desirable completion time for job j (there may exist a penalty for the late or for the early completion of that job). A job *preemption* might be allowed, i.e., it might be split into portions, each portion being assigned at a different time interval to the machine(s). A *(feasible) schedule* assigns each job j to the machine(s) at the specified time moment(s) no less than r_j with the total duration of p_j so that no two jobs are assigned to the machine at any time moment (i.e., the job execution intervals cannot overlap in time). A job is *late* (*on time*, respectively) if it is completed after (at or before, respectively) its due date.

In the single-machine scheduling problems, there is a single machine on which all the jobs are to be scheduled. The majority of single-machine single-criterion scheduling problems are NP-hard, although there are polynomially solvable cases as well. For instance, if the objective function is the maximum job completion time called the *makespan* and denoted by $C_{\max}$, then the problem of minimizing $C_{\max}$, commonly abbreviated by $1||C_{\max}$ according to the standard Graham's notation for scheduling problems, is straightforwardly solvable if each job j has a single parameter p_j (the processing time): schedule the jobs in any order without creating machine idle time before the first scheduled job and between any pair of jobs. It is very easy to see that this list scheduling algorithm gives an optimal solution. If each job j has also a release time r_j (the problem $1|r_j|C_{\max}$), then scheduling the jobs in any order may not be good, but still there is a very simple greedy way to arrange them optimally: just order the jobs with non-decreasing release times and iteratively assign the next job from the list to the machine at the completion time of the previously assigned job or at the release time of the former job, whichever magnitude is larger.

Minimizing the makespan becomes more complicated with even two machines or if each job j has an additional job parameter called the *delivery time* q_j, which is an extra amount of time needed for job j for its full completion *once* it is already completed on the machine (the delivery of each job is accomplished independently of the machine immediately after its completion on the machine). Thus, job j will take p_j time on the machine and then an additional time q_j for its full completion (during which another job might be assigned to the machine). Then the maximum job completion time in the schedule σ (the makespan) is:

$$C_{\max}(\sigma) = \max_{j \in \sigma} \left\{ s_j(\sigma) + p_j + q_j \right\}. \tag{1}$$

The objective is to find a feasible schedule in which the maximum job completion time is the minimum possible one.

If there are no job release times, i.e., all jobs are released simultaneously (the problem $1|q_j|C_{\max}$), then the makespan can be minimized by the well-known Jackson heuristic [5]: first arranging the jobs in a non-increasing order of their delivery times and then scheduling them without leaving machine idle times, similarly as we did for the above versions. With job release times, however, the problem $1|r_j, q_j|C_{\max}$ becomes strongly NP-hard. Besides the $C_{\max}$ criterion, there are a number of other commonly used objective functions for scheduling problems. For instance, if for every job j its due date d_j is given, then several objective criteria can be used to measure the solution quality.

The *lateness* of a job j in a schedule σ:

$$L_j(\sigma) = d_j - \left(s_j(\sigma) + p_j \right) \tag{2}$$

(note that $s_j(\sigma) + p_j$ is the completion time of job j in the schedule σ). One of the most commonly used due date oriented objective functions is the maximum job lateness

$$L_{\max}(\sigma) = \max_{j \in \sigma} L_j. \tag{3}$$

The objective is to find a feasible schedule σ in which the maximum job lateness $L_{\max}$ is the minimum possible one. This problem $1|r_j|L_{\max}$ is, in fact, equivalent to the abovementioned one $1|r_j, q_j|C_{max}$ with job delivery times, and hence, it is also strongly NP-hard [6].

Another common due date-oriented objective function is the number of *late* jobs (the ones completed after their due date)

$$\sum_{j \in \sigma} U_j(\sigma), \tag{4}$$

where $U_j(\sigma)$ is a 0–1 function taking the value 1 if job j is late in the schedule σ and the value 0 otherwise. The objective here is to find a feasible schedule with the minimum possible value $\sum_{j \in \sigma} U_j(\sigma)$, equivalently, one maximizing the throughput, i.e., the number of jobs completed by their due dates (this model is motivated by applications in real-time overloaded systems, where the job due dates are crucial in a way that if a job is late, then it might rather be postponed for an undefined period of time in favor of other jobs which might be completed on time). Similarly to the above problems, if all jobs are simultaneously released, then the problem

$1||\sum U_j$ is polynomially solvable (by the algorithm of Moore and Hodgson); however, with job release times, the problem $1|r_j|\sum U_j$ is again strongly NP-hard.

Hoogeveen [7] has considered the no machine idle time version in a bi-criteria setting. Instead of minimizing the lateness, he has introduced the so-called target start time s_j of a job j: s_j is the desirable starting time for job j, similarly as the due date d_j is the desirable completion time for the job j. Together with the minimization of the maximum job lateness, the minimization of the maximum job promptness (the difference between the target and real start times of that job) can be considered. The above reference gives an algorithm that finds a Pareto-optimal set of feasible solutions for this bi-criteria scheduling problem.

4. Basic multi-criteria scheduling problems

We can combine the objective functions described in the previous section and obtain the corresponding multi-criteria scheduling problems. We consider these multi-criteria problems from the point of view of multi-threshold optimization and Pareto-optimization approaches.

We start by considering a bi-criteria problem with two objective functions, $C_{\max}$ and $L_{\max}$ obtained from the single-criterion problems $1|r_j|C_{\max}$ and $1|r_j|L_{\max}$, respectively (note that in the first problem, no job delivery times are given).

With the Pareto-optimization approach, we need to solve two relevant problems: (1) among all feasible schedules with a given maximum job lateness, find one with the minimum makespan, and (2) vice versa, among all feasible schedules with a given makespan, find one with the minimum maximum job lateness. Both of these problems are strongly NP-hard [6].

With the multi-threshold (bi-threshold) optimization approach, we are given two threshold values A^1 and A^2 on the functions $C_{\max}$ and $L_{\max}$, respectively. We would like to know if there exists a feasible schedule σ such that

$$C_{\max}(\sigma) \leq A^1 \tag{5}$$

$$L_{\max}(\sigma) \leq A^2. \tag{6}$$

As to condition (5), let us first construct a feasible schedule σ' in which the jobs are arranged in a non-decreasing order of their release times and are scheduled in this order without leaving unavoidable machine idle time. Recall that the schedule σ' (obtained in this way in $O(n \log n)$ time) is optimal for the problem $1|r_j|C_{\max}$. Hence, if $C_{\max}(\sigma') > A^1$, then there exists no (bi-threshold optimal) schedule σ with $C_{\max}(\sigma) \leq A^1$ and we return a "no" answer. Otherwise, we know that there exists a feasible schedule σ' with $C_{\max}(\sigma) \leq A^1$. In fact, if $C_{\max}(\sigma') = A^1$, then there are many such feasible schedules (we may introduce idle time intervals of a required total length in the schedule σ arbitrarily between neighboring jobs in different ways obtaining different feasible schedules satisfying inequality (5)). Let us denote the set of these feasible schedules by S_{A^1}.

Now it remains to verify condition (6), i.e., we wish to know if, among all schedules from the set S_{A^1}, there is one satisfying condition (6). In general, it may take an exponential time to answer this question for an arbitrary value A^2 since the corresponding decision problem is NP-complete. At the same time, it also might be possible to obtain an answer in polynomial time, depending on the value of A^2. The easiest way is to construct a greedy solution σ'' to the problem obtained, for instance, by the earlier mentioned Jackson heuristic. It is well-known that the schedule σ'' minimizes the function $C_{\max}$. Hence, if $L_{\max}(\sigma'') \leq A^2$, then we return the schedule σ'' with a "yes" answer. Otherwise, the answer may be "yes" or may also be "no." In this case, we need more costly calculations to seek for a feasible schedule σ from the set S_{A^1} with $L_{\max}(\sigma) \leq A^2$. This may take an exponential time (as the second single-criterion problem $1|r_j|L_{\max}$ is NP-hard).

Combining the objective function $C_{\max}$ with $\sum_j U_j$, we obtain another bi-criteria problem from the single-criterion problems $1|r_j|C_{\max}$ and $1|r_j|\sum_j U_j$, respectively.

With the Pareto-optimization approach, we need to solve two relevant problems: (1) among all feasible schedules with a given maximum job lateness, find one with the minimum makespan, and (2) vice versa, among all feasible schedules with a given makespan, find one with the minimum number of late jobs. Both of these problems remain strongly NP-hard.

With the bi-threshold optimization approach, we are given two threshold values A^1 and A^3 on the functions $C_{\max}$ and $\sum_j U_j$, respectively. We would like to know if there exists a feasible schedule σ satisfying inequality (1) and the following inequality:

$$\sum_j U_j(\sigma) \leq A^3. \tag{7}$$

Condition (5) can be treated as above. As to condition (7), we need to verify if, among all schedules from the set S_{A^1}, there is one satisfying this condition. As for condition (6), in general, it may take an exponential time to verify condition (7) for an arbitrary value A^3, since the corresponding decision problem with a single objective function $\sum_j U_j$ is NP-complete [8]. But it again might be possible to obtain an answer in polynomial time. Instead of Jackson's heuristic that we used for condition (6), now we use an extended version of the algorithm of Moore and Hodgson for the problem $1||\sum U_j$. Recall that the latter algorithm is designed for simultaneously released jobs. It sorts all jobs in a non-decreasing order of their due dates and includes them in this order whenever the last included job completes by its due date. Otherwise, from the last block of the continuously scheduled jobs (there will be only one such block for simultaneously released jobs), it discards a longest job and repeats the same step until all jobs are considered in this way. Note that all the included jobs are completed on time. Finally, it adds the discarded jobs at the end of the resultant partial schedule in any order without leaving machine idle times (these jobs are late).

We modify the above algorithm by considering the jobs in the order as they are released, but order each group of currently released jobs similarly by non-decreasing due dates and accomplish the same steps for each such group of the already released jobs. Although the modified algorithm, in general, does not guarantee optimality, it may typically deliver a near-optimal solution to the version $1|r_j|\sum U_j$ with job release times. Let us denote the schedule delivered by the extended Moore and Hodgson algorithm by σ'''. It can be readily verified that the schedule σ''' minimizes the function $C_{\max}$. Hence, if $\sum_j U_j(\sigma''') \leq A^2$, then we return the schedule σ''' with a "yes" answer. Otherwise, the answer may be "yes" or may also be "no." In this case, we need more costly calculations to seek for a feasible schedule σ from the set S_{A^1} with $\sum_j U_j(\sigma) \leq A^2$, which, similarly as for the earlier bi-criteria problem, may take an exponential time.

Finally, combining all the three objective functions $C_{\max}$, $L_{\max}$, and $\sum_j U_j$, we obtain a more complicated three-criteria scheduling problem. Finding the Pareto-optimal set of feasible solutions obviously remains NP-hard. The by-threshold problem gets also less accessible but still more flexible than the Pareto-optimality version, again essentially depending on the threshold values. We again consider the three conditions (5), (6), and (7) that come from the corresponding single-criterion problems and the set of feasible schedules S_{A^1} yielded by inequality (1). Using the fact that both schedules σ'' and σ''' are from the set S_{A^1}, it will suffice to verify whether

$$\sum_j U_j(\sigma'') \leq A^3 \tag{8}$$

or

$$L_{\max}(\sigma''') \leq A^2. \tag{9}$$

Intuitively, it is clear that the closer is A^3 to n (the total number of jobs) and the larger is A^2, the more probable it is that these inequalities will hold. Hence, the by-threshold problem will be solved in $O(n \log n)$ time (remind that the time complexity of all the three heuristics that we use for the creation of the schedules σ', σ'', and σ''' is $O(n \log n)$). If any of the conditions (6), (7), (8), or (9) is not satisfied, then an implicit enumeration algorithm that generates feasible schedules respecting the thresholds A^2 and A^3 can be applied.

5. Conclusions

We have seen that a multi-threshold optimization problem may solve practical multi-criteria problems in polynomial time while delivering a solution with an acceptable quality for a given threshold vector, which reflects real needs of a particular real-life application. We have compared the multi-threshold optimization problem with the Pareto-optimization problem for three basic multi-criteria scheduling problems on a single machine. It is clear that, in many multi-criteria applications, a practitioner may not be interested in a Pareto-optimal set of feasible solutions: an analysis of the set of Pareto-optimal solutions containing all non-dominated feasible solutions might be beyond the interest and capacity of the practitioner. In practice, a feasible solution that attains some threshold value for each objective function is required. For instance, take an automobile manufacturing and the three objective functions $C_{\max}$, $L_{\max}$, and $\sum_j U_j$ considered in the previous section. Clearly, the manufacturer is interested in minimizing the total production time $C_{\max}$, whereas he imposes a maximum possible lateness in the production of each car (which might be far above the minimum possible lateness), and there is a maximum admissible number of cars whose production might be late and be delayed for an infinitive amount of time (according to the current demand on the product). Two heuristic algorithms that we have considered in the previous section, in practice, may well deliver such solutions while minimizing the total production time. It is well-known that Jackson's heuristic, in practice, delivers near-optimal solutions with a value of the objective function close to the optimum [9]. At the same time, if the threshold for the criterion $\sum_j U_j$ is not too small, the solution delivered by the heuristic may also satisfy the threshold condition for that criterion. In fact, it might be possible to combine Jackson's heuristic with Moore and Hodgson's one in such a way that the resultant heuristic would provide a solution with the desired thresholds for both objective functions with some high probability. The construction of such heuristics that deliver a solution respecting the threshold vector for two or more objective criteria is an interesting line for further research.

We have illustrated the multi-threshold optimization approach on a few single-machine scheduling problems, though the approach can obviously be applied, in general, for different kinds of multi-objective optimization problems.

Author details

Nodari Vakhania[1*] and Frank Werner[2]

1 Centro de Investigación en Ciencias, UAEM, Mexico

2 Fakultät für Mathematik, Otto-von-Guericke-Universität Magdeburg, Germany

*Address all correspondence to: nodari@uaem.mx

References

[1] T'kindt V, Billaut J-C. Multicriteria Scheduling: Theory, Models and Algorithms. 2nd ed. Springer; 2006

[2] T'kindt V, Billaut J-C. Multicriteria scheduling: A survey, RAIRO. Operations Research. 2001;**35**:143-163

[3] Nagar A, Haddock J, Heragu S. Multiple and bicriteria scheduling: A literature review. European Journal of Operational Research. 1995;**81**(1): 88-104

[4] Parveen S, Ullah H. Review on job-shop and flow-shop scheduling using multi criteria decision making. Journal of Mechanical Engineering. 2010;**41**(2):130-146

[5] Jackson JR. Scheduling a Production Line to Minimize the Maximum Tardiness. Management Science Research Project. Los Angeles, CA: University of California; 1955

[6] Vakhania N. Scheduling a single machine with primary and secondary objectives. Algorithms. 2018;**11**:80. DOI: 10.3390/a11060080

[7] Hoogeveen H. Single-machine bicriteria scheduling [Ph. D. Thesis]. Amsterdam: CWI; 1992

[8] Garey MR, Johnson DS. Computers and Intractability: A Guide to the Theory of NP–Completeness. San Francisco: Freeman; 1979

[9] Vakhania N, Perez D, Carballo L. Theoretical expectation versus practical performance of Jackson's heuristic. Mathematical Problems in Engineering. 2015;**2015**:484671. DOI: 10.1155/2015/484671

Section 3

Applications and Overviews

Chapter 4

Overview of Multi-Objective Optimization Approaches in Construction Project Management

Ibraheem Alothaimeen and David Arditi

Abstract

The difficulties that are met in construction projects include budget issues, contractual time constraints, complying with sustainability rating systems, meeting local building codes, and achieving the desired quality level, to name but a few. Construction researchers have proposed and construction practitioners have used optimization strategies to meet various objectives over the years. They started out by optimizing one objective at a time (e.g., minimizing construction cost) while disregarding others. Because the objectives of construction projects often conflict with each other, single-objective optimization does not offer practical solutions as optimizing one objective would often adversely affect the other objectives that are not being optimized. They then experimented with multi-objective optimization. The many multi-objective optimization approaches that they used have their own advantages and drawbacks when used in some scenarios with different sets of objectives. In this chapter, a review is presented of 16 multi-objective optimization approaches used in 55 research studies performed in the construction industry and that were published in the period 2012–2016. The discussion highlights the strengths and weaknesses of these approaches when used in different scenarios.

Keywords: construction project management, multi-objective optimization, evolutionary algorithms, swarm intelligence algorithms, analytic network process, nature-based algorithms, Hungarian algorithm, mixed-integer nonlinear programming, hybrid approaches

1. Introduction

The main objective of the construction industry is to directly and indirectly provide people's daily needs. Mostly, a construction project involves the use of different resources (e.g., machinery, materials, manpower, etc.) to produce the final product (e.g., a building, a bridge, a water distribution system, etc.) that serves the targeted users' needs. The difficulties that are met in construction projects include budget limitations, contractual time constraints, safety and health issues, sustainability ratings, local building codes, the desired level of quality, to name but a few. Consequently, a construction project has multiple objectives including maximum productivity, minimum cost, minimum duration, specified quality, safety, and sustainability. Making decisions is difficult when one wants to reach the optimal solution for a combination of objectives.

Construction practitioners have been using single-objective optimization strategies to meet the desired level of construction objectives. However, because the multiple objectives of construction projects often conflict with each other, single-objective optimization does not offer practical solutions, as optimizing one objective would often adversely affect the other objectives that are not being optimized. As a result, some projects fail to meet some of the objectives. In order to avoid such failures, researchers have developed tools that can help efficiently manage construction projects and achieve the required objectives. These tools include many multi-objective optimization approaches, each of which has its own advantages and drawbacks when used in some scenarios with different sets of objectives.

A review is presented in this chapter of the various multi-objective optimization approaches used in recent studies in the construction industry to highlight the strengths and weaknesses of these approaches when used in different scenarios.

2. Overview

A total of 55 studies that applied multi-objective optimization methods in the construction industry are reviewed in this chapter. To avoid overlapping and redundancy of reviews with Evins' work [1], the review in this chapter includes only the recent studies which were published in the period late 2012 to early 2016. Evins [1] covered the period of 1990 to late 2012 and conducted a review of the studies that applied optimization methods in sustainable building design.

The 55 studies are reviewed relative to (1) the optimization method, (2) the project phase, (3) the optimization problem, (4) the type and number of targeted objectives, (5) the example used to test a model, and (6) the comparison with other methods when applicable.

The number of optimization methods found in the review of the 55 papers was 16. These 16 methods and their usage frequency are presented in **Figure 1**, which shows that NSGA-II is the most used method (14 times) followed by a hybrid method (12 times) which pairs two or more methods for the optimization process. The acronyms in this figure are spelled out in **Table 1**.

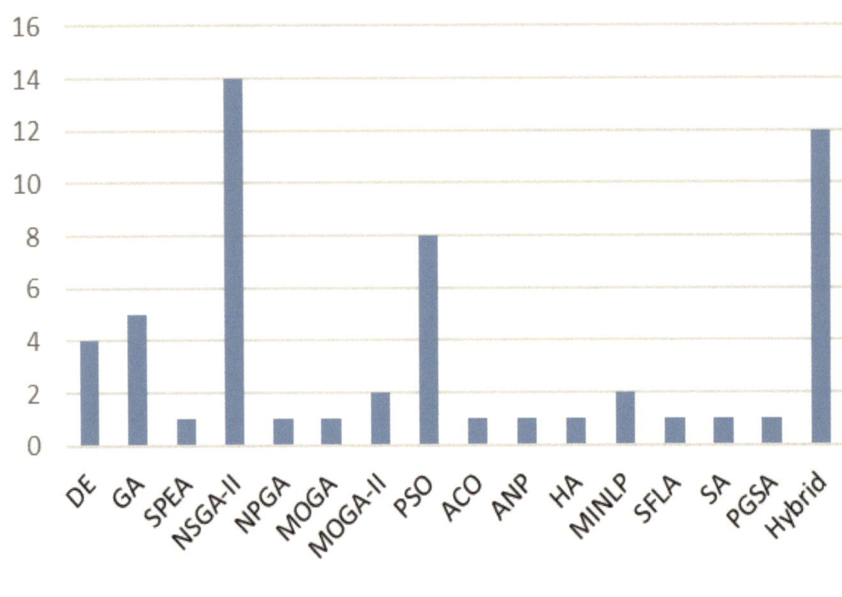

Figure 1.
Frequency of methods used in literature.

Optimization method	Number of objectives					
	2	3	4	5	6	7
Genetic algorithms (GA)	2	3	—	—	—	—
Differential evolution (DE)	1	3	—	—	—	—
Strength Pareto evolutionary algorithm (SPEA)	—	1	—	—	—	—
Non-dominated sorting genetic algorithm-II (NSGA-II)	8	6	—	—	—	—
Niched Pareto genetic algorithm (NPGA)	—	1	—	—	—	—
Multi-objective genetic algorithm (MOGA)	1	—	—	—	1	1
Particle swarm optimization (PSO)	3	3	—	2	—	—
Ant colony optimization (ACO)	1	—	—	—	—	—
Analytic network process (ANP)	—	—	1	—	—	—
Shuffled frog-leaping algorithm (SFLA)	—	1	—	—	—	—
Simulated annealing algorithm (SA)	1	—	—	—	—	—
Plant growth simulation algorithm (PGSA)	1	—	—	—	—	—
Hungarian algorithm (HA)	1	—	—	—	—	—
Mixed-integer nonlinear programming (MINLP)	2	—	—	—	—	—
Hybrid methods	6	6	—	—	—	—
Total (56 methods)	27	24	1	2	1	1

Table 1.
Number of objectives used in the literature.

These optimization methods were used to tackle different numbers of objectives at a time. The number of objectives that was simultaneously optimized ranged between 2 and 7. The most common number of objectives in a study was 2 or 3 objectives (27 and 24 times, respectively) distributed by methods as shown in **Table 1**. The least common number of objectives considered in a study was 4, 6, and 7 (one time each). It should be noted that one of the 55 papers used two optimization methods, i.e., NSGA-II and PSO. Therefore, the total number of methods used in the 55 papers is 56.

As expected, the large majority of the studies optimized two or three objectives that concern most practitioners. The number of times the objectives were used is presented in **Table 2**. Among the objectives used in the 55 papers, cost was the mostly optimized, accounting for 93% (51 times) of the total number of studies, duration was the second most optimized objective accounting for 42% (23 times), and the energy and environment category was the third most optimized with 31% (17 times). The rating system score was used only 3 times, i.e., in only 5% of the studies, which represents the least optimized objective.

3. Multi-objective optimization methods used in recent construction-related studies

3.1 Genetic algorithms (GA)

GA is one of the popular evolutionary algorithms used by researchers. GA uses the concept of chromosomes to present the possible solutions in these chromosomes' strings [2]. The different aspects of each solution are positioned into the slots

Objective	Number of times objective used in studies
Cost	51
Duration	23
Quality	7
Resources	7
Energy and environment	17
Thermal	13
Safety	6
Rating system score	3
Other	23

Table 2.
Number of times the objectives were used in the 55 studies.

which form the string [3]. A new set of solutions are found by the crossover between two strings (parent strings), and the new strings (children) will inherit the best features of the parent strings.

In construction-related fields, GA has been applied in many multi-objective optimization problems. For example:

- GA was used to improve sustainability in housing units. Karatas and El-Rayes [4] used GA in a single-family housing unit to optimize operational environmental performance, social quality of life, and life cycle cost. They used 33 decision variables in the model and computed in 47.5 hours 210 near-optimal solutions within a large search space of configurations and decisions (more than 2.6 quadrillion).

- GA was used to solve conflicting objectives in construction scheduling. For instance, Agrama [5] used GA to optimize building schedules. The author analyzed a 5-storey building and used nine scenarios for the weights of three objectives: project duration, total actual crews, and total interruptions for all activities. The model was implemented in Excel (Evolver) and solved by GA. In addition, it was found that the model performs consistently and can be used with both the critical path and line of balance methods. Moreover, the results obtained were identical to those in the literature but required less time and effort. Alternatively, Aziz et al. [6] introduced a method that combines CPM with GA to optimize the utilization of resources for mega construction projects in terms of time, cost, and quality. An 18-activity schedule was tested using the proposed method. To avoid complexity, the five decision variables which were construction materials, crew formation, crew overtime policy, machinery efficiency, and contractor class were all combined into a single decision variable called resource utilization. In this test, 305 optimal solutions were identified. Additionally, the results showed that the model outperformed the approach used by Feng et al. [7] with the same case example.

- GA was used in managing site operations. For example, in material logistics, Said and El-Rayes [8] presented an example of a 10-storey building consisting of 107 activities with four temporary facilities. The aim of the model was to minimize total construction logistics costs (Eq. (1)) and minimize project schedule criticality (Eq. (2)).

$$Min\ TLC = OC + FC + SC + LC \quad (1)$$

$$Min\ SCI = \frac{1}{N'} \times \sum_{i=1}^{N'} CI_i = \frac{1}{N'} \times \sum_{i=1}^{N'} \frac{SS_i - ES_i}{TF_i} \quad (2)$$

where, TLC = total logistics costs; OC = ordering cost; FC = financing cost; SC = stock-out cost; LC = layout cost; SCI = schedule criticality index; N' = number of noncritical activities; CI_i = criticality index of activity i; SS_i = scheduled start time of activity i; ES_i = early start time of activity i; and TF_i = total float of activity i.

Because the search space is large and the problem is complex, the authors justified the use of a GA model that involves 152 decision variables and 462 constraints. The model generated 361 optimal solutions. For equipment management problems, Xu et al. [9] proposed dynamic programming-based GA because they believed it would be capable of solving this type of problem more efficiently than traditional methods. The goal of the method was to minimize the project's total cost and maximize equipment operations such that in case of equipment failure there would be an equipment available. Moreover, to make the method more reliable, the failure rate of the equipment was considered a fuzzy variable. An actual hydropower project in China was selected to test the model. Under the same environment, the proposed algorithm performed better in searching than the standard GA.

In summary, there is evidence that GA can optimize different objectives in the construction industry in the field of scheduling, sustainability, and site operation.

3.2 Differential evolution (DE)

The DE approach is efficient and has low algorithmic complexity. There is also some evidence of its effectiveness in tackling problems of continuous optimization with different types of constraints and functions [10]. The members of the population in DE use floating-points which identify each member's direction and distance [11]. Therefore, the main concept behind the DE approach is that it creates a new population member with a vector that has the difference between two members' vectors; that process is done by the mutation and crossover processes [12].

DE has proved its effectiveness in complex planning and scheduling problems by optimizing cost and time in addition to quality, environmental impact, or resources. For example:

- Narayanan and Suribabu [13] applied DE to assist contractors in optimizing their plans for subcontracting in terms of cost, time and quality. To examine the model, they used a 7-activity and an 18-activity project. By comparison, the DE model generated better solutions than ant colony optimization (ACO) for cost in the first case, and for cost and time in the second case.

- Alternatively, Cheng and Tran [14] used a two-phase DE model on a 37-activity warehouse project to minimize total project cost and duration, while accounting for resource constraints. In the first phase, a multiple objective DE model was used to find the optimal tradeoff between time and cost in construction activities. Based on the solution obtained in the first phase, the best schedule was found within resource constraints in the second phase. A comparison of the results showed that the developed model outperformed three evolutionary algorithms: DE, particle swarm optimization (PSO) and NSGA-II.

- Subsequently, Cheng and Tran [15] proposed opposition-based multi-objective DE. The aim was to optimize construction products in terms of cost, time and environmental impact. The model used opposition-based learning to increase precision and convergence speed. A tunnel project consisting of 25 activities was used to test the model. The proposed model was superior compared to NSGA-II, PSO, and DE algorithms. The exact approach also outperformed these algorithms in a similar study conducted by Cheng and Tran [16].

- The goal of the Cheng and Tran [16] study was to minimize project time (Eq. (3)), project cost (Eq. (4)), and the utilization of resources (Eqs. (5) and (6)) in overtime shifts.

$$Min\ T = \sum_{n=1}^{l} T_n^{S_n} = \mathrm{Max}_{\forall n}(\mathrm{ES_n} + D_n) \tag{3}$$

$$Min\ \mathrm{C} = \sum_{n=1}^{N} Cost_n^{S_n} \tag{4}$$

$$Min\ LHEN = LHE + LHN(1 + W)\ if\ \mathrm{SS} = 3\ (three - shift\ system) \tag{5}$$

$$Min\ LHNE = LHE\ if\ \mathrm{SS} = 2\ (\mathrm{two} - shift\ system) \tag{6}$$

where in Eq. (3), $T_n^{S_n}$ is the duration of the activity $n\{n = 1, 2, \ldots, l\}$ on the critical path for a specific option of resources (S_n); l is the total number of critical activities on a specific critical path; ES_n is the earliest start of activity n; D_n is the duration of activity n. In Eq. (4), $Cost_n^{S_n}$ is the total cost which includes direct and indirect cost of activity n for a specific option of resources (S_n); N is the total number of activities. In Eqs. (5) and (6), *LHE* is the total number of evening shift work hours and *LHN* is the total number of night shift work hours. Because risks faced in night shiftwork are typically higher than in other shifts, *W* is the defined weight that represents the relative importance of minimizing *LHN*.

A 15-activity and a 60-activity project were used to test the model. In just one run, the model was capable of finding Pareto-optimal solutions to solve the objectives of the problem.

It can be concluded that the DE algorithm is capable of optimizing several objectives of time, cost, resource utilization, and environmental impact. Moreover, as DE and its variations successfully optimized those objectives, they also surpassed ACO, PSO, and NSGA-II in construction scheduling optimization.

3.3 Strength Pareto evolutionary algorithm (SPEA)

SPEA works by archiving the non-dominated solutions found in the Pareto-front at every iteration. Then, based on the number of solutions it dominates, each solution in the archive is ranked with a strength rate [10, 17].

In dealing with scheduling problems, SPEA was proposed by Elazouni and Abido [18] to optimize the three conflicting objectives of maximizing profit and minimizing required finance and resource idle days. The study used two examples from the literature to test the efficiency and scalability of the model. In the first example, the model was tested for its effectiveness in solving a 9-activity project. The model confirmed its robustness by achieving 50 identical solutions. By searching these solutions using a fuzzy based method, the top ones were selected. In the second example, an 18-activity project was used to assess the model's scalability.

Four solutions (maximum profit, minimum finance, minimum resource idle days, and the top compromised solution) were drawn from the 48 solutions obtained in the Pareto-optimal front. Clustering the Pareto solutions set was used to keep it within a manageable size. Nevertheless, because of the clustering, this method may result in the loss of some extreme Pareto solutions.

By optimizing the construction objectives of profit and resources, SPEA has verified its efficiency in the scheduling field. However, the clustering method proposed by Elazouni and Abido [18] should be avoided when using SPEA in order to avoid the elimination of some extreme Pareto solutions. New clustering approaches should be explored in upcoming studies.

3.4 Non-dominated sorting genetic algorithm-II (NSGA-II)

One of the most powerful tools of genetic algorithms is NSGA-II. It uses the non-dominated sorting for the solutions in the population. The non-dominated solutions are ranked at every iteration, and are excluded from the population in every iteration afterwards. In addition, in each ranked-solution set, the solutions are compared to each other by their crowding formation. In the crowding step, the position of a single solution is measured by its distance from the adjacent solutions' points, and based on its distance, the solution is assigned with a rank, as the best ranks start from the shortest distance to the longest one [10].

NSGA-II has been used to solve multi-objective problems aimed at the optimality of energy consumption and sustainability in buildings. For instance:

- Eliades et al. [19] used NSGA-II to optimally select the installation locations for indoor air quality sensors, in terms of number of sensors, and average and worst-case impact damage while considering the building's usage in the parameters. A simple 5-room building and a 14-room house were studied to illustrate the performance of the proposed model, with 5 and 2310 contamination scenarios, respectively. Grid and random sampling were used to construct the contamination scenarios, and the multi-zone building program CONTAM simulated them.

- In zero-energy-building (ZEB), Hamdy et al. [20] used a modified version of NSGA-II to find solutions for the optimal cost and nearly zero energy building performance with respect to the guidelines of European directives for the energy performance of buildings. Due to the large number of combinations, the solution space was divided into three stages. The total number of combinations (179, 712) in the first stage were searched in 800 runs.

- Huws and Jankovic [21] took into account future weather changes that could affect retrofitting strategies. These weather changes may eventually unsettle the performance of zero-carbon buildings by increasing the carbon emissions or cost, or in some cases a combination of these may create thermal discomfort. For that reason and to achieve optimal solutions for retrofit, environmental, social, and economic constraints were considered in optimizing the objectives of minimizing cost, CO_2, and thermal discomfort. A simple 60 m^2 box model was created using the DesignBuilder program. DesignBuilder and JEPlus were used to perform the optimization process. NSGA-II within JEPlus was used for its capability of searching a large solutions space, and to avoid being stuck in a local suboptimum. The results indicated that there is an applicable alternative for both current and future weather.

- In sustainability for low-income housing, Marzouk and Metawie [22] incorporated NSGA-II with BIM to assist the Egyptian government find solutions that best optimize those objectives. The BIM model was created using Revit. The model was defined based on the quantities and properties of the materials extracted from the BIM model. These quantities helped to find the different solutions in terms of project cost, duration and maximum LEED points. Construction productivity and cost were determined using a 44-activitiy low-income housing building. Moreover, LEED points were calculated through five credits chosen from the materials and resources category.

- Kasinalis et al. [23] studied the improvement of indoor environment while reducing the energy consumption in climate adaptive building shells, and quantified the impact of using seasonal adaptation façade on those objectives. The example of an office zone model was used to evaluate the approach. The combination of daylight and energy simulations were utilized with NSGA-II to perform multi-objective optimization on that example. The optimization process considered six design parameters for the façade. The results showed that using a seasonal adaptation façade with these parameters is more efficient than a non-adaptive façade, since it can save up to 18% of energy consumption and enhance the quality of the indoor environment.

- Inyim et al. [24] approached the problem of building components and material selection by using (SimulEICon) a BIM tool that simulates the environmental impact in buildings. The optimization process of time, cost and CO_2 emissions was performed by NSGA-II. The case study was an actual zero net energy house. The model considered 16 activities and 185 building components. It was found that some of the combinations of components suggested by SimulEICon matched the original component combinations used in the existing house. However, SimulEICon lacked the ability to account for more than three objectives.

- Carreras et al. [25] introduced an approach for selecting the thickness of insulation material for building shells. The objective of the study was to select the best option for the insulation that optimizes the costs (Eq. (7)) and environmental impacts (Eq. (8)) associated with energy consumption.

$$Min\ Cost_{total} = Cost_{cub} + Cost_{elec\_n} \tag{7}$$

$$Min\ Imp_{total} = Imp_{cub} + Imp_{elec} \tag{8}$$

where $Cost_{total}$ is the total cost, $Cost_{cub}$ is the cost of the materials used; $Cost_{elec\_n}$ is the cost of the electricity consumed over the operational phase (n years); Imp_{total} is the total environmental impact; Imp_{cub} is the total impact of the materials used; and Imp_{elec} is the impact of the consumed electricity over the operational phase.

The authors used the example of a cubicle without insulation to compare the different results collected from using two cases of insulation. In the first case, similar thicknesses were used over the cubicle, while in the second case, different thicknesses were considered. Three materials were considered in the insulation selection process (polyurethane, mineral wool, and polystyrene). From the results, the polyurethane insulation was the least costly solution, whereas the optimal environmental impact solution was mineral wool insulation. The proposed methodology could improve the costs and environmental impacts by almost 40% when compared to a non-insulated cubicle.

Site operations and planning problems were also tackled using NSGA-II. For instance:

- Fallah-Mehdipour et al. [26] applied NSGA-II to solve two tradeoff problems, time-cost and time-cost-quality, respectively. To validate the proposed method, an 18-activity and a 7-activity work schedule were utilized. Additionally, multi-objective PSO was applied. The results showed that NSGA-II was superior to multi-objective PSO.

- In managing and storing materials in a construction site, Said and El-Rayes [27] presented an automated module, which imports its data from BIM files and historical schedule data. A module in the system was named construction logistics planning (CLP) and aimed to minimize the cost of logistics and the criticality of the schedule. These objectives were optimized by tackling four decision variables using NSGA-II. An application model of a 10-storey building project was used to apply the optimization process. The automated system generated better results compared to using CLP alone. A total of 361 optimal solutions were produced within 65 hours. Unlike CLP, which considered the utilization of exterior site space and disregarded the interior one, the system generated the solutions accounting for both spaces.

- In site operations, Parente et al. [28] proposed NSGA-II to optimize the allocation of compaction equipment within the criteria of cost and time associated with earthworks in large infrastructure projects. Additionally, linear programming was used for the allocation of the remaining equipment such as trucks and excavators. The proposed method which uses an actual construction site as a case study proved to be superior to the S-metric selection evolutionary algorithm as well as manual allocation.

NSGA-II was used to find solutions in problems involving upgrade plans for water networks and slum areas. For example:

- Creaco et al. [29] divided the construction phases of a water network upgrade into four phases, considering the different phases of upgrades to the water network in a 100-year plan of possible upgrades. NSGA-II was used with a model of six network nodes and eight pipe laying locations to find the optimal solutions within the two objective functions: maximizing the minimum pressure and minimizing the cost, while the pipe diameters are acting as the decision variables. The proposed approach provided better results than the studies that used single phasing, by giving the optimal solution for maintaining the water distribution and pressure quality through the time of upgrade phases. In a similar study, Creaco et al. [30] proposed the use of NSGA-II while considering an additional factor to the study, which was the uncertainty of water demand. The authors determined the uncertainty using a probabilistic approach. Based on an example with 26 network nodes and 31 pipe laying locations. The probabilistic approach was compared with the deterministic approach used by Creaco et al. [29]. The results revealed that the solutions obtained by the probabilistic approach had higher costs than the solutions of the deterministic approach, especially in the first phase. However, the probabilistic solutions generated better results in terms of costs when the comparison was about the worst-case scenario.

- In uneven ground levels of slum areas, El-Anwar and Abdel Aziz [31] used an example of nine-zone slum area with a population of 2770 families to select the optimal upgrade plan. The optimization process involved three objectives: maximization of benefit of proposed upgrading projects, minimization of costs and socioeconomic disruption for the families. Due to its superiority over other GAs in solving multi-objective problems, NSGA-II was selected to solve the problem in which it generated 2000 solutions in less than 1 minute. Nevertheless, the time schedules module was not included in the model hence affecting its robustness.

- Brownlee and Wright [32] analyzed the relationship between design objectives and the effectiveness of design variables on the design objectives by using NSGA-II. They sorted the objectives by simple ranking. The approach was performed on a five-zone building with only two design objectives. The objectives to be minimized were total annual energy use and capital cost, and the design variables were 52 in total. Forty-nine solutions were generated using NSGA-II. However, the proposed approach failed to discriminate the distance variables which are the variables that measure the sets from the true Pareto-optimal set from the floating variables which are the variables that have no effect on the objective function.

As the above-cited studies show, the NSGA-II proved its capability in optimizing for scheduling, urban planning, infrastructure, sustainability, energy and environmental design, and resource management. In addition to its superiority over other GAs, NSGA-II has also outperformed other methods in some fields. One of those is the multi-objective PSO applied to scheduling problems.

3.5 Niched Pareto genetic algorithm (NPGA)

The tournament selection among a population's individuals and Pareto dominance are the two basic ideas of NPGA's process. The selection process is based on the dominance of two randomly selected individuals from the population. To determine which individual of these two is dominant over the other, another set of individuals are picked and used to go against the two competing individuals, to examine the level of the two competing individuals in dominating each individual of the set. The winning criterion is defined by Pareto-front dominance. Therefore, one of the two competing individuals is selected if the other is dominated by one of the individuals in the set [33, 34].

Kim et al. [35] used NPGA to optimize cost, time and resource utilization. They optimized the three objectives at the same time. To test the performance of the method, they conducted two case studies. The first case had 11 activities, and measured the method's efficiency in solving the tradeoff problem between cost and time. In addition to the objectives in the first case, the second case extended the examination of the approach by including the resource-leveling index as an objective. The results showed that this method could provide decision makers with different solutions to enable them selecting the one that meets their preferences.

3.6 Multi-objective genetic algorithm (MOGA)

MOGA is an advanced version of traditional GA. The difference between MOGA and GA is the individual fitness assignment, while the remaining steps are followed as in GA. In MOGA, ranking is assigned for each individual in the population. The rank is assigned based on individual's dominance, if the individual is not dominated by another individual in the population then it is assigned with the rank of one. But if an individual is dominated by other individuals then it is assigned with a rank corresponding to the total number of dominating individuals plus one [36].

- MOGA has shown its capabilities in achieving optimal structural design. For example, Richardson et al. [37] tackled the design problem of an x-bracing structural system for a building façade. Minimizing the cost of the bracing connections and the effectiveness of the bracings were the objectives under the multi-objective topology optimization process (Eq. (9)).

$$\min_{x} f(x) = (f_1, f_2) \tag{9}$$

where f_1 is the cost objective function expressed in Eq. (10), x is the variable vector of length n, and f_2 is the relative tier deflection objective function expressed in Eq. (11).

$$Min\ \mathrm{f}_1 = \sum_{i=1}^{n} a_i.x_i \tag{10}$$

$$Min f_2 = \max\left\{\frac{|d_1|}{h_1}, \frac{|d_2 - d_1|}{h_2}, \frac{|d_3 - d_2|}{h_3}\right\} \tag{11}$$

where a_i is a weighting coefficient related to the grouping of components based on symmetry; x_i is the topology variable associated with bracing(s) i; h_j is the height of tier j; and d_j is the measured deflection of tier j from rest position.

While the constraints change as the design progresses, the proposed approach dynamically adapts to those constraints. Museum façades were picked to test the performance of the optimization method.

- In reducing the energy consumed and environmental impact in buildings, Baglivo et al. [38] have used an improved version of MOGA (MOGA-II) on combinations of sustainable building materials for external walls of zero energy buildings, to achieve the best optimal solutions in balancing the life cycle cost and energy consumption. The materials were tested according to their thermal characteristics based on the Mediterranean climate. The assessment of material combinations was carried on six thermal-related objectives. The study concluded that the best selection of materials for external walls was by placing the insulation coating on the external side of the wall, while placing the high internal capacity material on the interior side. Similarly, Baglivo et al. [39] have conducted a study that added one more objective to the same six objectives.

3.7 Particle swarm optimization (PSO)

The pattern of flocking birds and fish was the inspiration of PSO. In PSO, a set of solutions is called swarm, while a solution is called particle [26]. The particles are positioned in a D-dimensional search space. In each step, every particle changes its velocity to move toward the best solution and toward the global best solution [40].

Different issues of construction engineering and management were tackled by PSO. Some studies proposed PSO to solve site planning problems. For instance:

- Xu and Li [41] proposed permutation-based PSO to solve the planning problem of a dynamic construction site layout, in which ordinal numbers assigned to the particles were used to present the potential solutions. The objectives considered in the problem were the layout cost and the environmental and safety accidents. Since the study accounted for uncertainty, fuzzy random variables were included in the model. The model used the example of 14 temporary facilities in a hydropower project to evaluate its efficiency. The proposed approach proved to be more realistic than existing traditional approaches.

- Xu and Song [42] approached the problem of unequal-area departments in dynamic temporary facility layout using position-based adaptive PSO. By using the facilities' coordinates as base for its model, the optimization process aimed at minimizing the total distance between adjacent facilities and the resulting costs associated with rearrangement and transportation, in which the transportation costs were considered as fuzzy random variables. The modified PSO was evaluated through a case study of a large-scale hydropower construction project. The proposed method showed better performance in obtaining optimal solutions when compared to standard PSO and GA.

- Li et al. [43] proposed a modified PSO to achieve optimal solutions for dynamic construction site layout and security planning. The study approached the problem using the Stackelberg Game theory, in which the construction manager (the leader) must set up the layout and secure the facilities, then the attacker (the follower) has to create the maximum possible economic damage to the facilities. Bi-level multi-objective PSO was proposed to solve the problem. The method was implemented in a hydropower construction project to test its performance. The proposed method outperformed GA in achieving optimal solutions.

PSO has also been used in tackling different objectives in the maintenance of deteriorating structures. For example:

- Yang et al. [44] approached the problem of life cycle maintenance planning for deteriorating bridges using PSO with Monte Carlo simulation (MCS). Cost, safety and condition levels were the main objectives in the maintenance problem. Uncertainties in the maintenance cost, work effects of maintenance, and the structure' deterioration rate were also accounted for in the study. Parallel programming was used to minimize the computing time to solve the problem. Yang et al. [44] considered three paradigms in the programming process, namely master-slave, island, and diffusion. In each paradigm, the computers have a different set up to run MCS in parallel. From the analysis, the island paradigm surpassed the other two in terms of solution quality. By comparison, the multi-objective PSO algorithm outperformed NSGA-II.

- Chiu and Lin [45] proposed PSO to achieve the optimal strategies in maintaining reinforced concrete buildings. The authors considered five objectives in the study, which are life cycle cost, failure possibility, spalling possibility, maintenance rationality, and maintenance times. Assessment models of probabilistic effects were employed to observe the effects of maintenance strategies on the damage index. The four processes of analysis of deterioration, assessment of seismic performance, forming maintenance strategies, and multi-objective optimization were performed in the proposed maintenance strategy. The evaluation was completed using a case study of a four-story reinforced concrete school building.

Some researchers used PSO to tackle different design objectives and constraints to achieve optimal sustainable design solutions. For instance:

- Decker et al. [46] have proposed a PSO algorithm to reach better design solutions in timber buildings. In addition to architectural, energy and environmental constraints, the study added structural constraints. The optimization process was in terms of energy needs, thermal discomfort, floor vibration, CO_2 emissions, and embodied energy. To minimize computing time, the simulation model was transformed using a metamodeling procedure. A three-story office building was used as a case study to validate the proposed approach.

- Chou and Le [47] used PSO in combination with MCS to attain the optimal solutions for building designs in terms of minimizing duration (Eq. (12)), cost (Eq. (13)), and CO_2 emissions (Eq. (14)).

$$Min\ F_{dur} = \mathrm{ES}_{\mathrm{fin}} + \sum_{i=1}^{n} ES_i \tag{12}$$

$$Min\ F_{cost} = \sum_{i=1}^{n} W_i.COST_i \tag{13}$$

$$Min\ F_{CO_2} = \sum_{i=1}^{n} W_i.FC_i \tag{14}$$

where F_{dur}, F_{cost}, and F_{CO2} represent project duration, project cost, and CO_2 emissions, respectively; ES_{fin} is the early start of the finish activity; ES_i is the early start of activity i; $COST_i$ is the unit cost of activity i; n is the number of activities; and FC_i is the amount of CO_2 emitted to complete a unit of work of activity i.

In addition to PSO, a probabilistic method was applied to handle the uncertainties associated with the objectives of the study. The case study of a 12-activity roadway pavement project was used to evaluate the performance of the proposed method.

In sum, PSO proved its effectiveness in tackling the multi-objective optimization problems in different construction engineering and management areas such as site planning, maintenance of a structure, and sustainability issues. It was found that PSO's performance was superior compared to traditional approaches such as GA and advanced approaches such as NSGA-II.

3.8 Ant colony optimization (ACO)

The stimulus in discovering the ACO algorithm was the movement of ants and their trails of pheromones when searching for food. In the ACO process, each solution is connected to a route that is searched by an ant. Each solution's quality is evaluated by the quantity of pheromones that were deposited on the route by ants. The amount of pheromone left on a route indicates the closeness to the optimal solution. The chance of finding the shortest route increases for an ant as the amount of pheromone on a route increases [48].

The proximity and number of construction facilities and other resources on a construction site could contribute to an increase in cost and safety issues. Ning and Lam [49] developed a modified ACO model to tackle safety and cost problems within a site layout of irregular shape. The model was aimed to minimizing safety/

environmental concerns by reducing the occurrence of accidents (Eq. (15)) as well as minimizing the total handling cost between facilities by reducing the cost associated with resource exchanges among facilities (Eq. (16)).

$$Min\, f_1 = \min \sum_{i=1}^{n} \sum_{j=1}^{n} \sum_{l=1}^{n} \sum_{k=1}^{n} S_{ij} d_{kl} x_{ik} x_{jl} \tag{15}$$

$$Min\, f_2 = \min \sum_{i=1}^{n} \sum_{j=1}^{n} \sum_{l=1}^{n} \sum_{k=1}^{n} C_{ij} d_{kl} x_{ik} x_{jl} \tag{16}$$

where S_{ij} is the closeness relationship value for safety/environmental concerns between facilities i and j; C_{ij} is the total closeness relationship value for the total handling cost between facilities i and j; d_{kl} is the distance between facilities k and l; x_{ik} means when facility i is assigned to location k; and x_{jl} means when facility j is assigned to location l.

The optimization process started by dividing the site layout into a grid. The grid units were selected based upon the dimensions of the facilities. Then, the ACO model was used to assign the different facilities on the site grid. To test the soundness of the model, a residential project composed of four buildings was selected. The proposed grid strategy reduced the complexity of the computational process.

3.9 Analytic network process (ANP)

Like the analytic hierarchy process, decision makers utilize ANP to solve multi-criteria decision problems. The AHP uses a one-way top-down hierarchal process for its components such as goals, criteria, and alternatives [50]. The ANP which is a generalized version of AHP uses a network for some problems when their components have interdependencies between them. The flow in the ANP's network is open and allows any component to interact with another regardless of their levels, which is not possible in AHP [51].

Liang and Wey [52] proposed an ANP model to optimally select government projects by accounting for the limitation of resources along with uncertainties and socioeconomic factors. In order to test the model's effectiveness, seven projects in a nation-wide highway improvement project were used as an example. In the example, construction costs were determined by probability distributions and seven criteria were used to evaluate the projects. Moreover, since the model involves the use of multiple criteria, ANP was combined with MCS to make the selection of projects based on the solutions achieved by solving the multi-objective problems. ANP ranking was used to rank each project based on its value of priority among other projects. A cost-benefit approach was used to optimize the selection of projects based on the existing budget plan and the allocation of remaining budget to fund a project in full. The four objectives within these two problems were minimization of cost (Eq. (17)) and the number of project managers (Eq. (18)), and the maximization of project ranking (Eq. (19)) and the number of completed projects (Eq. (20)).

$$Minimize\ modified\ mean\ absolute\ deviation\ of\ cost = Min \frac{\sum_{i=1}^{n} \sum_{j=1}^{m} y_{ij}^{+}}{nm} \tag{17}$$

$$Minimize\ number\ of\ project\ managers = Min \sum_{i=1}^{n} PMNO_i x_i \tag{18}$$

$$Maximize\ project\ ranking = Max \sum_{i=1}^{n} RANK_i x_i \quad (19)$$

$$Maximize\ number\ of\ completed\ projects = Max \sum_{i=1}^{n} x_i \quad (20)$$

where n is the number of projects; m is the number of scenarios; y_{ij}^{+} is the positive deviation of the cost of the scenario from the expected cost of the project; $RANK_i$ is the ranking given to project i based on the ANP computation; x_i is a binary variable which has a value of 1 if project i is selected, and 0 otherwise; and $PMNO_i$ is the number of project managers needed to complete project i.

3.10 Shuffled frog-leaping algorithm (SFLA)

The SFLA idea is based on frogs' behavior in their search to locate the largest quantities of food [53]. A single solution is represented by one frog [54, 55]. The frogs are divided into groups (memeplexes). Each memeplex of frogs performs a local search, and every frog has an idea which is affected by other frogs' ideas to improve the quality of the local search [56]. A shuffling process is performed to allow the memeplexes in exchanging information between them and create new memeplexes to ultimately improve their quality of search [53, 54, 56].

Improving the quality of the final product with limited resources is the ultimate goal of construction managers and planners. Time, cost, and resources play important roles in achieving this goal. Ashuri and Tavakolan [57] concurrently optimized three objectives: the duration expressed by sum of the durations of activities on the critical path, the project cost including direct and indirect costs, and resource allocation variations expressed in Eq. (21).

$$Min\ (SSR) = min \left(\sum_{k=1}^{TD} \sum_{n=1}^{S} R_{n,d}^2 \right) \quad (21)$$

where $R_{n,d}$ is the number of the nth resource with n = 1, 2, … , S that is planned for use in day d with d = 1, 2, … , TD of the project duration.

To solve these problems, they used the SFLA model. In order to find feasible solutions to the problem at hand, the model accounts for the reallocation of resources and for activity interruptions and splitting. In addition, the authors made use of the advantages of PSO and the shuffling complex evolution algorithm, which helped the model achieve better solutions and converge more rapidly. A 7-activity and an 18-activity project were utilized to assess the efficiency of the model. Delphi was the coding environment for the model. Due to the complexity of the problem, the solutions obtained were near-optimal. However, the proposed model generated better solutions than other algorithms used prior to the study.

3.11 Simulated annealing algorithm (SA)

SA inherits its method from the movements of atoms within a material during the process of heating and then slowly cooling down [58]. In the optimization problem, the physical system's characteristics resemble the actual annealing process [10]. Talbi [10] listed the characteristics of physical annealing with their corresponding characteristics of the optimization problem. In physical annealing, temperature and speed of cooling down play important roles on the strength of metals. Deficiencies (metastable states) occur when cooling down speed is fast or the temperature at the start is not high enough [59]. That means carefully setting up the temperature and cooling down speed is essential in escaping the local optimum —metastable state in physical annealing—and reaching the global optimum. A solution that is generated after an iteration is used, if feasible, to generate a new solution, but if the solution is infeasible, it is accepted only if it meets the probability criterion [10, 60]. The probability increases in obtaining an optimal or near-optimal solution when the annealing is slowed down [61].

To optimally design and construct a water distribution network, Marques et al. [62] proposed a model that used the SA algorithm combined with the EPANET hydraulic simulator. The objective was to minimize the cost of construction and operation including the initial and future costs, and to minimize violations in pressure as expressed in Eq. (22).

$$Min\ TPV = \sum_{s=1}^{NS}\sum_{t=1}^{NTI}\sum_{d=1}^{NDC}\sum_{n=1}^{NN} \max\{0; (Pdes_{\min,n,d} - P_{n,d,t,s})\} \tag{22}$$

where TPV is the total pressure violations; NS is the number of scenarios; NTI is the number of periods into which the planning horizon is subdivided; NDC is the number of demand conditions considered for the design; NN is the number of nodes; $Pdes_{min,n,d}$ is the minimum desirable pressure at node n for demand condition d; and $P_{n,d,t,s}$ is the pressure at node n at demand condition d for time interval t and in scenario s.

Eight scenarios were accounted for varying between three possible patterns of growth in the area: expansion, no expansion, and depopulation in a 60-year period. They split the 60-year duration of the plan into 320-year stages, and structured them into a decision tree to show the probability of the paths in each scenario. They used a 17-node distribution network to illustrate the model's efficiency. The decision variables included cost, diameters of pipes (eight diameters were considered), and carbon emissions produced during construction and operation (in terms of tons). The value of the objective function was not noticeably affected by the decision variable of carbon emission costs.

3.12 Plant growth simulation algorithm (PGSA)

The PGSA imitates the growth process of trees. The model's formulation for the optimization process in PGSA is based on the growth of plants [63]. It begins at the root then moves toward the light source (global optimum solution) to grow the branches [64]. A probability model is employed to form new branches which are used to guide the objective function toward the optimal solution [65].

To better minimize the losses and costs caused by an attack to the construction site and to increase the safety precautions to counter these attacks, Li et al. [66] used a bi-level model. The objectives of reducing attack-related cost and increasing facility productivity were considered at the upper level, in which the secured facilities were constrained by cost. The attacker, on the other hand, has the objective of reducing facility productivity, which is considered in the lower level. The formulation of the objective functions is as follows:

$$Max_{z_j} \mathrm{D} = \sum_{j=1}^{N}\sum_{i=1,\, i\neq j}^{N}\sum_{k=1}^{5} \theta_{ij} p_k \mu_{ijk}^{r} d_{ij} s_j + \sum_{j=1}^{N}\sum_{i=1,\, i\neq j}^{N} \theta_{ij} d_{ij} z_j + \sum_{j=1}^{N}\sum_{i=1,\, i\neq j}^{N} \theta_{ij} d_{ij} (1 - s_j)(1 - z_j) \tag{23}$$

where D is the resource supply rate; z_j is 1 when facility j is secured and 0 otherwise; s_j is 1 when facility j is attacked and 0 otherwise; θ_{ij} is the weight of demand's importance; $0 \leq \theta_{ij} \leq 1$; d_{ij} is 1 when demand of facility i is served by facility j and 0 otherwise; p_k is the occurrence probabilities of the kth degree attack, $k \in \{1, \ldots, 5\}$; and μ^r_{ijk} is the mean value of the fill rate of facility j to facility i when facility j is attacked.

$$\text{Min}_{z_j} C = \sum_{j=1}^{N} \sum_{k=1}^{5} p_k \mu^c_{jk} s_j + \sum_{j=1}^{N} M_j z_j \tag{24}$$

where C is the economic loss of defender; M_j is the cost of securing facility j*; and* μ^c_{jk} is the mean value of the economic loss when facility j is attacked.

Because integer programming made the problem complicated, the authors proposed PGSA. The model was applied on an actual hydropower project. Fifty runs were executed to achieve the optimal solution in less than 4 minutes. Even though the proposed model efficiently solved the problem, it did not top the list of algorithms. This study was the first study to apply PGSA on the problem of construction site security.

3.13 Hungarian algorithm (HA)

The Hungarian algorithm is a modified form of the primal-dual algorithm that is used to solve network flows. In assignment problems, the Hungarian algorithm changes the weights in a matrix to locate the optimal assignment. Eventually, a new matrix is obtained in which the optimal assignment is identified [67].

El-Anwar and Chen [68] proposed a modified Hungarian algorithm to solve post-disaster temporary housing problems. They considered the problem as an integer problem. An earthquake simulation example was used to examine the model's efficiency. The number of decision variables was determined by multiplying the housing alternatives (178) with the number displaced families (5000). Throughout the 13 temporary housing problems, a varying number of decision variables were considered. In terms of the running time, the Hungarian algorithm has shown superiority over integer programming. In the example, the running time for integer programming increased exponentially as the number of decision variables increased, and ran out of memory in case more than 24,000 decision variables were used. The Hungarian algorithm, on the other hand, solved all the problems with the maximum number of decision variables (890,000).

3.14 Mixed-integer nonlinear programming (MINLP)

MINLP is an optimization problem in which the variables are constrained to continuous (e.g., costs, dimensions, mass, etc.) and integer values (typically binary 0 and 1), and the solution space and the objective functions are represented by nonlinear functions [69–71]. To solve complex problems that involve nonlinearity and mixed-integers, MINLP utilizes the combination of mixed-integer programming (MIP) and nonlinear programming (NLP) [72]. Thus, in solving MINLP problems, the approach is not considered a direct problem solver. The methods used to solve MINLP optimization problems include: branch and bound method, benders decomposition, and outer approximation algorithm [73].

- Fan and Xia [74] used MINLP to reduce energy consumption in residential buildings. The objectives of the study were to increase the energy savings and economic benefits within budget limitations. The example of a 69-year old house was used to test the model, in which the retrofitting plan included the insulation materials for the roof and external walls, windows, and the installation of solar panels. The model proved to be effective in minimizing the energy consumed by the building; from the results obtained, in a 10-year period, the house could save around 288.44 MWh.

- Karmellos et al. [75] also used MINLP to optimize the energy used by a building. The minimization of energy consumption every year and the cost of investments were the two main objectives in the optimization process. To test the model's soundness, the energy consumption in two houses was investigated. The first case involved a new house located in the UK while the second case was an existing house located in Greece. Fifty-four decision variables were accounted for in the model, which represented different building components including electrical appliances, building envelope, and lighting and energy systems. The model was effective in solving the optimization problem of and building energy. It was found that energy consumption goes down when investments in energy efficiency are increased.

3.15 Hybrid approaches

One way in approaching complex optimization problems is to combine two or more techniques together in order to overcome the deficiencies that one or some of them may possess. This approach could affect the overall quality of the solution in an optimization problem. The hybridization of methods has shown its efficacy in accomplishing optimization quality in construction. Hybrid methods have different operational characteristics in tackling optimization problems. While some hybrid methods work by carrying the entire solution process as a single novel technique, others work in tandem whereby one method works on some steps of the solution process and the other steps are completed by another method.

NSGA-II was hybridized with other approaches to solve optimization problems in construction planning, scheduling, energy conservation, transportation, and environmental design. For example:

- Mungle et al. [76] used fuzzy clustering-based genetic algorithm (FCGA) to find optimal solutions for the trade-off problem of time, cost and quality within the construction tasks. The method hybridized the fuzzy clustering approach with NSGA-II. In addition, AHP was utilized to measure the construction activities' quality. To evaluate the model's efficiency, a highway construction project was selected as an example. The authors used the example in three cases with different number of activities, i.e., eighteen, twelve, and seven-activity networks, in which the proposed approach was compared to other methods. The results of the comparison showed that FCGA surpassed MOPSO, MOGA and SPEA-II in terms of diversity as well as the speed and degree of convergence.

- Monghasemi et al. [77] proposed an approach that combines NSGA-II with MOGA to solve a discrete problem of cost, time, and quality in construction project scheduling. An 18-activity highway construction project was used to examine the proposed model. MOGA was utilized to search the large size of 3.6 billion solutions and obtain near true optimal solutions. Shannon's entropy method was used to assign normalized weights to the three objectives in the obtained solutions. These weights were used to rank the solutions by performing the evidential reasoning method, which assist decision makers in assessing each solution in terms of performance.

- Brownlee and Wright [78] proposed modified approaches of NSGA-II on a simulation-based optimization problem for building energy. The minimization of energy usage and construction cost were the two objectives in the optimization process. The aim of the study was to find an approach that surpasses the basic NSGA-II in terms of convergence rate and solution quality. The study used a middle floor from a commercial office building in three different cities to test the proposed model. The authors merged NSGA-II with two other approaches, namely radial basis function networks and deterministic infeasibility sorting. These approaches enabled the model to prevent the elimination of infeasible solutions and to keep them in the population. The objectives were represented by 50 decision variables (30 continuous, 8 integers and 12 categorical) and 18 inequality constraints. Moreover, the optimization runs were limited to 5000 completed within almost a day by six parallel simulations. The model was found superior to the basic NSGA-II in two of the three building examples.

- Xu et al. [79] proposed a multi-objective bi-level PSO (MOBLPSO) to optimize the minimum cost network flow of construction material transportation in terms of duration and cost. In the upper level of the model, the time to transport materials in addition to direct costs were minimized by the contractor by selecting the most convenient routes for transporting materials. Depending on the decisions made in the model's upper level, every transporter's flow of material in those routes were considered by the transportation manager to reduce the costs of transportation. Because of the complexity of the problem the PSO approach was hybridized with two other methods, one in each level. In the upper level, PSO was integrated with Pareto Archived Evolution Strategy (PAES) to keep the best position for the solutions up to date. In the lower level, PSO used passive congregation to prevent the convergence from happening too early. The case of an actual hydropower construction project was utilized to examine the model's soundness. The model outperformed multi-objective bi-level genetic algorithms (MOBLGA) and the multi-objective bi-level simulated annealing algorithm (MOBLSA).

- Xu et al. [80] conducted a similar study to the one mentioned above, but in this study the cost and duration of transportation were considered as fuzzy random variables. A fuzzy random simulation-based constraint checking procedure was coupled with MOBLPSO to solve the transportation assignment problem which was used to control the flow of materials within a given period. The road network of an existing hydropower project was used for the evaluation of the model. With accounting for uncertainties, the model showed its efficiency and capability of solving the transportation problem.

- Zhang et al. [81] proposed immune genetic PSO (IGPSO) which couples immune genetic algorithm with PSO. The approach was used to tackle the trade-off problem of time-cost-quality in construction, and accounting for bonus and penalty. The hybrid method in the research obtained its characteristics from three methods: (1) from the immune algorithm, whereby the hybrid method inherits the immune selection and the memory recognition; (2) from the genetic algorithm, which implements mutation and crossover; and (3) by limiting the particles' maximum velocity using the constriction factor in PSO, which speeds up the convergence in initial steps. In addition, the study used a PERT network instead of CPM for the schedule. The model was applied on the 19 activities of a three-floor office building, and proved its effectiveness in solving the trade-off problem.

- In the trade-off problem, some researchers used the double-loop technique, in which the internal loop executes the simulation, while the external loop carries out the optimization process. However, this technique uses MCS and can sometimes take days to finish the process. Therefore, Yang et al. [82] proposed a procedure that combines the double-loop into one, and used MCS and support vector regression (SVR) with PSO to expedite the process of obtaining optimal solutions for the time-cost trade-off problem. MCS was set to assess the initial solutions' objective values, and a decision function gained by SVR promptly assesses the solutions obtained by PSO for their objective values. SVR's direct assessment contributed in shortening the search time of MCS. To test the model, an 18-activity project was utilized as an example. The results obtained showed that the proposed method was superior compared to the methods that used the double loop.

- Futrell et al. [83] used PSO coupled with Hooke Jeeves and the generic optimization program (GenOpt) to optimize the performance of daylighting and thermal systems in buildings. Hooke Jeeves was utilized to fine-tune the best solution found in the PSO algorithm by locally searching it. The case study of a classroom design was utilized to evaluate the proposed approach. The classroom was tested on windows facing north, south, west, and east. It was found that there was no significant conflict between the optimization objectives when the windows were facing south, west, or east, but there was a significant conflict between those objectives in the case of windows facing north.

- Yahya and Saka [84] used multi-objective artificial bee colony (ABC) with the Levy flights algorithm to find the ideal layout for a construction site. Levy flight which uses a random walk pattern searches food locations found by ABC to locate new solutions. The objective functions of the study were the reduction of the facilities' total handling cost, and minimization of environmental and safety risks. Two practical study cases were used to assess the proposed model. The first case was a residential project consisting of four high-rise buildings, and the second case was a three-floor private hospital. The first case which was a dynamic site layout was taken from Ning et al.'s [85] study, in which they applied a modified ACO. From the results, the model succeeded in optimizing the site layout problems. By comparison, the method proposed by Yahya and Saka [84] surpassed the plain-ABC and the modified ACO used by Ning et al. [85].

- Tran et al. [86] tackled the trade-off problem of time, cost, and quality using the combination of multi-objective ABC with DE. DE was included to use its crossover mutation operators to optimize the stages of exploration and exploitation. A study case of a construction project consisting of 60 activities was used to test the model. The result proved the model's efficacy in the trade-off problem. The approach was also compared against four other approaches that were used to solve the trade-off problem. The proposed method outperformed multi-objective ABC, multi-objective DE, multi-objective PSO, and NSGA-II.

- Marzouk et al. [87] presented a hybrid approach that combined ACO with system dynamics to optimize the selection of sustainable materials. The aim of the study was the maximization of LEED credits and the minimization of cost. The authors employed a study case of a two-floor residential building to validate the efficacy of the model. From the achieved results, the model proved its capability in accomplishing the two objectives of the problem.

- In building maintenance planning, Wang and Xia [88] used a predictive control model and DE algorithm to achieve the optimal retrofitting plan that lowers energy consumption. The study's first objective aimed at increasing a project's internal rate of return. The study's second objective was to increase energy savings while accounting for the sustainability period. The authors tackled the optimization of the maintenance plan in two instances. They started by solving the optimization problem without the assumption of uncertainties. They then solved the problem with uncertainties, in which the predictive control model was utilized. To check the approach's validity, a case study that involved the retrofitting of an office building consisting of 50 stories was considered. The results showed that the proposed approach was effective in finding the optimal maintenance plan.

The complexity of the problems in construction projects makes objective optimization usually difficult using a single approach. Hybrid techniques are effective and useful in generating optimal solutions in complex optimization problems. In some studies, these hybrid methods have outperformed some methods in their basic and variant forms. In scheduling for example, they were superior to multi-objective PSO, multi-objective ABC, multi-objective DE, MOGA, SPEA-II, and NSGA-II. In material logistics, they surpassed multi-objective bi-level GA and multi-objective bi-level SA. In site planning, they outperformed the basic form of ABC and one of the ACO variants. Finally, in sustainability, they were superior to NSGA-II.

4. Conclusion

This review included 55 papers that were published in refereed journals and conference proceedings published in the years 2012–2016. The authors of these papers conducted studies using various multi-objective optimization methods in the construction industry. There were 16 methods used in these studies in which some of the authors justify their picks on multiple factors (e.g., construction project type, project size, number of objectives, number of constraints, convergence rate, problem complexity such as constraints' nonlinearity with discontinuity and continuity, etc.). Moreover, some methods were found to be more efficient than others in some studies. For example, in water network planning, Creaco et al. [30] showed that their NSGA-II using a probabilistic approach was superior to NSGA-II used by Creaco et al. [29] in an earlier study in which they used a deterministic approach. The GA proposed by Aziz et al. [6] in a scheduling problem outperformed the GA utilized by Feng et al. [7] for the same case study. Fallah-Mehdipour et al. [26] concluded that NSGA-II has performed better than multi-objective PSO in solving a scheduling problem. Most of the time, it is difficult to guarantee the performance of a method until it is compared with another method.

The most common number of objectives used in the literature is two and three. As expected, cost and duration were the most targeted objectives as cost and duration are important objectives for all construction practitioners. The quality objective has also drawn the interest of researchers as they sometimes include it in trade-off problems with cost and/or duration. However, quality has not been optimized in any other set of objectives than three-objective optimization problems. The energy and environment category is an important candidate in the optimization process, as it came after cost and duration objectives based on the number of times it was optimized. That may be the result of efforts to optimally construct sustainable buildings and lower the depletion of natural resources.

Among the multi-objective methods used in the literature, NSGA-II was the most used method. NSGA-II has proven its capability in solving optimization problems in different fields of construction. In addition to its popularity among researchers, NSGA-II has many advantages that make it suitable for many types of optimization problems such as obtaining diverse solutions in Pareto-front, low computational complexity, solving problems that involve nonlinearity and discontinuity.

Author details

Ibraheem Alothaimeen and David Arditi*
Illinois Institute of Technology, Chicago, IL, USA

*Address all correspondence to: arditi@iit.edu

References

[1] Evins R. A review of computational optimisation methods applied to sustainable building design. Renewable and Sustainable Energy Reviews. 2013; **22**:230-245. DOI: 10.1016/j.rser.2013.02.004

[2] Whitley D. A genetic algorithm tutorial. Statistics and Computing. 1994; **4**(2):65-85. DOI: 10.1007/BF00175354

[3] Sivanandam S, Deepa S. Genetic algorithm optimization problems. In: Introduction to Genetic Algorithms. Berlin, Heidelberg: Springer Science & Business Media; 2008. pp. 165-209. DOI: 10.1007/978-3-540-73190-0_7

[4] Karatas A, El-Rayes K. Optimizing tradeoffs among housing sustainability objectives. Automation in Construction. 2015;**53**:83-94. DOI: 10.1016/j.autcon.2015.02.010

[5] Agrama FA. Multi-objective genetic optimization for scheduling a multi-storey building. Automation in Construction. 2014;**44**:119-128. DOI: 10.1016/j.autcon.2014.04.005

[6] Aziz RF, Hafez SM, Abuel-Magd YR. Smart optimization for mega construction projects using artificial intelligence. Alexandria Engineering Journal. 2014;**53**(3):591-606. DOI: 10.1016/j.aej.2014.05.003

[7] Feng CW, Liu L, Burns SA. Using genetic algorithms to solve construction time-cost trade-off problems. Journal of Computing in Civil Engineering. 1997; **11**(3):184-189. DOI: 10.1061/(ASCE)0887-3801(1997)11:3(184)

[8] Said H, El-Rayes K. Optimal utilization of interior building spaces for material procurement and storage in congested construction sites. Automation in Construction. 2013;**31**: 292-306. DOI: 10.1016/j.autcon.2012.12.010

[9] Xu J, Meng J, Zeng Z, Wu S, Shen M. Resource sharing-based multi objective multistage construction equipment allocation under fuzzy environment. Journal of Construction Engineering and Management. 2013;**139**(2):161-173. DOI: 10.1061/(ASCE)CO.1943-7862.0000593

[10] Talbi EG. Metaheuristics: From Design to Implementation. Hoboken, NJ: John Wiley & Sons; 2009. p. 500. DOI: 10.1002/9780470496916

[11] Onwubolu GC, Babu BV. New Optimization Techniques in Engineering. Berlin, Heidelberg: Springer; 2004. p. 712. DOI: 10.1007/978-3-540-39930-8

[12] Boussaïd I, Lepagnot J, Siarry P. A survey on optimization metaheuristics. Information Sciences. 2013;**237**:82-117. DOI: 10.1016/j.ins.2013.02.041

[13] Narayanan AS, Suribabu CR. Multi-objective optimization of construction project time-cost-quality trade-off using differential evolution algorithm. Jordan Journal of Civil Engineering. 2014;**8**(4): 375-392

[14] Cheng MY, Tran DH. Two-phase differential evolution for the multiobjective optimization of time-cost tradeoffs in resource-constrained construction projects. IEEE Transactions on Engineering Management. 2014;**61**(3):450-461. DOI: 10.1109/TEM.2014.2327512

[15] Cheng MY, Tran DH. Opposition-based multiple-objective differential evolution to solve the time-cost-environment impact trade-off problem in construction projects. Journal of Computing in Civil Engineering. 2015; **29**(5):04014074. DOI: 10.1061/(ASCE)CP.1943-5487.0000386

[16] Cheng MY, Tran DH. Opposition-based multiple objective differential

evolution (OMODE) for optimizing work shift schedules. Automation in Construction. 2015;**55**:1-14. DOI: 10.1016/j.autcon.2015.03.021

[17] Simon D. Evolutionary Optimization Algorithms. Hoboken, NJ: John Wiley & Sons; 2013. p. 772

[18] Elazouni A, Abido MA. Enhanced trade-off of construction projects: Finance-resource-profit. Journal of Construction Engineering and Management. 2014;**140**(9):04014043. DOI: 10.1061/(ASCE)CO.1943-7862.0000880

[19] Eliades DG, Michaelides MP, Panayiotou CG, Polycarpou MM. Security-oriented sensor placement in intelligent buildings. Building and Environment. 2013;**63**:114-121. DOI: 10.1016/j.buildenv.2013.02.006

[20] Hamdy M, Hasan A, Siren K. A multi-stage optimization method for cost-optimal and nearly-zero-energy building solutions in line with the EPBD-recast 2010. Energy and Buildings. 2013;**56**(1):189-203. DOI: 10.1016/j.enbuild.2012.08.023

[21] Huws H, Jankovic L. A method for zero carbon design using multi-objective optimisation. In: Proceedings of the 1st International Conference on Zero Carbon Buildings Today and in the Future; 11-12 September 2014. Birmingham, United Kingdom: Birmingham City University; 2014. pp. 99-106

[22] Marzouk M, Metawie M. Framework for sustainable low-income housing projects in Egypt. In: Proceedings of the 2014 International Conference on Computing in Civil and Building Engineering; 23-25 June 2014; Orlando, Florida, United States. 2014. pp. 1960-1968

[23] Kasinalis C, Loonen RCGM, Cóstola D, Hensen JLM. Framework for assessing the performance potential of seasonally adaptable facades using multi-objective optimization. Energy and Buildings. 2014;**79**:106-113. DOI: 10.1016/j.enbuild.2014.04.045

[24] Inyim P, Rivera J, Zhu Y. Integration of building information modeling and economic and environmental impact analysis to support sustainable building design. Journal of Management in Engineering. 2015;**31**(1):A4014002. DOI: 10.1061/(ASCE)ME.1943-5479.0000308

[25] Carreras J, Boer D, Guillén-Gosálbez G, Cabeza LF, Medrano M, Jiménez L. Multi-objective optimization of thermal modelled cubicles considering the total cost and life cycle environmental impact. Energy and Buildings. 2015;**88**: 335-346. DOI: 10.1016/j.enbuild.2014.12.007

[26] Fallah-Mehdipour E, Haddad OB, Rezapour Tabari MM, Mariño MA. Extraction of decision alternatives in construction management projects: Application and adaptation of NSGA-II and MOPSO. Expert Systems With Applications. 2012;**39**(3): 2794-2803. DOI: 10.1016/j.eswa.2011.08.139

[27] Said H, El-Rayes K. Automated multi-objective construction logistics optimization system. Automation in Construction. 2014;**43**:110-122. DOI: 10.1016/j.autcon.2014.03.017

[28] Parente M, Cortez P, Gomes Correia A. An evolutionary multi-objective optimization system for earthworks. Expert Systems With Applications. 2015;**42**(19):6674-6685. DOI: 10.1016/j.eswa.2015.04.051

[29] Creaco E, Franchini M, Walski TM. Network design through the phasing of construction approach. Procedia Engineering. 2014;**89**:823-830. DOI: 10.1016/j.proeng.2014.11.513

[30] Creaco E, Franchini M, Walski TM. Taking account of uncertainty in demand growth when phasing the construction of a water distribution network. Journal of Water Resources Planning and Management. 2014; **141**(2):04014049. DOI: 10.1061/(ASCE) WR.1943-5452.0000441

[31] El-Anwar O, Abdel Aziz T. Integrated urban-construction planning framework for slum upgrading projects. Journal of Construction Engineering and Management. 2014;**140**(4):B4014003. DOI: 10.1061/(ASCE) CO.1943-7862.0000672

[32] Brownlee AEI, Wright JA. Solution analysis in multi-objective optimization. In: First Building Simulation and Optimization Conference; 10-11 September 2012; Loughborough, United Kingdom. 2012. pp. 317-324

[33] Zitzler E, Thiele L. Multiobjective evolutionary algorithms: A comparative case study and the strength Pareto approach. IEEE Transactions on Evolutionary Computation. 1999;**3**(4): 257-271. DOI: 10.1109/4235.797969

[34] Dias AHF, De Vasconcelos JA. Multiobjective genetic algorithms applied to solve optimization problems. IEEE Transactions on Magnetics. 2002; **38**(2):1133-1136. DOI: 10.1109/ 20.996290

[35] Kim K, Walewski J, Cho YK. Multiobjective construction schedule optimization using modified niched Pareto genetic algorithm. Journal of Management in Engineering. 2015; **32**(2):04015038. DOI: 10.1061/(ASCE) ME.1943-5479.0000374

[36] Awad M, Khanna R. Efficient Learning Machines: Theories, Concepts, and Applications for Engineers and System Designers. Berkeley, CA: Apress; 2015. p. 268. DOI: 10.1007/978-1-4302-5990-9

[37] Richardson JN, Nordenson G, Laberenne R, Filomeno Coelho R, Adriaenssens S. Flexible optimum design of a bracing system for façade design using multiobjective genetic algorithms. Automation in Construction. 2013;**32**:80-87. DOI: 10.1016/j.autcon.2012.12.018

[38] Baglivo C, Congedo PM, Fazio A, Laforgia D. Multi-objective optimization analysis for high efficiency external walls of zero energy buildings (ZEB) in the Mediterranean climate. Energy and Buildings. 2014;**84**:483-492. DOI: 10.1016/j.enbuild.2014.08.043

[39] Baglivo C, Congedo PM, Fazio A. Multi-criteria optimization analysis of external walls according to ITACA protocol for zero energy buildings in the Mediterranean climate. Building and Environment. 2014;**82**:467-480. DOI: 10.1016/j.buildenv.2014.09.019

[40] Van den Bergh F, Engelbrecht AP. A cooperative approach to particle swarm optimization. IEEE Transactions on Evolutionary Computation. 2004; **8**(3):225-239. DOI: 10.1109/ TEVC.2004.826069

[41] Xu J, Li Z. Multi-objective dynamic construction site layout planning in fuzzy random environment. Automation in Construction. 2012;**27**: 155-169. DOI: 10.1016/j.autcon.2012. 05.017

[42] Xu J, Song X. Multi-objective dynamic layout problem for temporary construction facilities with unequal-area departments under fuzzy random environment. Knowledge-Based Systems. 2015;**81**:30-45. DOI: 10.1016/j. knosys.2015.02.001

[43] Li Z, Shen W, Xu J, Lev B. Bilevel and multi-objective dynamic construction site layout and security planning. Automation in Construction. 2015;**57**:1-16. DOI: 10.1016/j.autcon. 2015.04.011

[44] Yang IT, Hsieh YM, Kung LO. Parallel computing platform for multiobjective simulation optimization of bridge maintenance planning. Journal of Construction Engineering and Management. 2012;**138**(2):215-226. DOI: 10.1061/(ASCE)CO.1943-7862.0000421

[45] Chiu CK, Lin YF. Multi-objective decision-making supporting system of maintenance strategies for deteriorating reinforced concrete buildings. Automation in Construction. 2014;**39**: 15-31. DOI: 10.1016/j.autcon.2013.11.005

[46] Decker SA, Ndiaye A, Brangeon B, Sempey A, Galimard P, Pauly M, et al. Design of multi-story timber building using mutli-objective particle swarm optimization. In: World Conference on Timber Engineering; 10-14 August 2014; Québec, Canada. 2014. p. 2

[47] Chou JS, Le TS. Probabilistic multiobjective optimization of sustainable engineering design. KSCE Journal of Civil Engineering. 2014; **18**(4):853-864. DOI: 10.1007/s12205-014-0373-x

[48] Parpinelli RS, Lopes HS, Freitas AA. Data mining with an ant colony optimization algorithm. IEEE Transactions on Evolutionary Computation. 2002;**6**(4):321-332. DOI: 10.1109/TEVC.2002.802452

[49] Ning X, Lam KC. Cost-safety trade-off in unequal-area construction site layout planning. Automation in Construction. 2013;**32**:96-103. DOI: 10.1016/j.autcon.2013.01.011

[50] Saaty TL. Decision making with the analytic hierarchy process. International Journal of Services Sciences. 2008;**1**(1): 83-98. DOI: 10.1504/IJSSCI.2008.017590

[51] Saaty TL, Vargas L. The analytic network process. In: Decision Making with the Analytic Network Process. Boston, MA: Springer. International Series in Operations Research & Management Science. 2006;**95**:1-26. DOI: 10.1007/0-387-33987-6_1

[52] Liang S, Wey WM. Resource allocation and uncertainty in transportation infrastructure planning: A study of highway improvement program in Taiwan. Habitat International. 2013;**39**:128-136. DOI: 10.1016/j.habitatint.2012.11.004

[53] Samuel GG, Rajan CCA. A modified shuffled frog leaping algorithm for long-term generation maintenance scheduling. In: Proceedings of the Third International Conference on Soft Computing for Problem Solving; 26-28 December 2013; Roorkee, India. New Delhi, India: Springer; 2014. pp. 11-24. DOI: 10.1007/978-81-322-1771-8_2

[54] Elbeltagi E, Hegazy T, Grierson D. A modified shuffled frog-leaping optimization algorithm: Applications to project management. Structure and Infrastructure Engineering. 2007;**3**(1): 53-60. DOI: 10.1080/15732470500254535

[55] Lu K, Ting L, Keming W, Hanbing Z, Makoto T, Bin Y. An improved shuffled frog-leaping algorithm for flexible job shop scheduling problem. Algorithms. 2015;**8**(1):19-31. DOI: 10.3390/a8010019

[56] Eusuff M, Lansey K, Pasha F. Shuffled frog-leaping algorithm: A memetic meta-heuristic for discrete optimization. Engineering Optimization. 2006;**38**(2):129-154. DOI: 10.1080/03052150500384759

[57] Ashuri B, Tavakolan M. Shuffled frog-leaping model for solving time-cost-resource optimization problems in construction project planning. Journal of Computing in Civil Engineering. 2015;**29**(1):04014026. DOI: 10.1061/(ASCE)CP.1943-5487.0000315

[58] Lamberti L. An efficient simulated annealing algorithm for design optimization of truss structures. Computers and Structures. 2008; **86**(19):1936-1953. DOI: 10.1016/j.compstruc.2008.02.004

[59] Kibria A, Castillo-Villar KK, Millwater H. Minimizing the discrepancy between simulated and historical failures in turbine engines: A simulation-based optimization method. Mathematical Problems in Engineering. 2015;**2015**(2):1-11. DOI: 10.1155/2015/813565

[60] Basu M. A simulated annealing-based goal-attainment method for economic emission load dispatch of fixed head hydrothermal power systems. International Journal of Electrical Power & Energy Systems. 2005;**27**(2):147-153. DOI: 10.1016/j.ijepes.2004.09.004

[61] Zhao X. Simulated annealing algorithm with adaptive neighborhood. Applied Soft Computing. 2011;**11**(2): 1827-1836. DOI: 10.1016/j.asoc.2010.05.029

[62] Marques J, Cunha M, Savić DA. Multi-objective optimization of water distribution systems based on a real options approach. Environmental Modelling and Software. 2015;**63**:1-13. DOI: 10.1016/j.envsoft.2014.09.014

[63] Wang C, Cheng HZ, Hu ZC, Wang Y. Distribution system optimization planning based on plant growth simulation algorithm. Journal of Shanghai Jiaotong University (Science). 2008;**13**(4):462-467. DOI: 10.1007/s12204-008-0462-4

[64] Rao PVVR, Sivanagaraju S. Radial distribution network reconfiguration for loss reduction and load balancing using plant growth simulation algorithm. International Journal on Electrical Engineering and Informatics. 2010; **2**(4):266-277. DOI: 10.15676/ijeei.2010.2.4.2

[65] Guney K, Durmus A, Basbug S. A plant growth simulation algorithm for pattern nulling of linear antenna arrays by amplitude control. Progress in Electromagnetics Research B. 2009;**17**: 69-84. DOI: 10.2528/PIERB09061709

[66] Li Z, Xu J, Shen W, Lev B, Lei X. Bilevel multi-objective construction site security planning with twofold random phenomenon. Journal of Industrial and Management Optimization. 2015;**11**(2): 595-617. DOI: 10.3934/jimo.2015.11.595

[67] Brauner N, Echahed R, Finke G, Gregor H, Prost F. Specializing narrowing for timetable generation: A case study. In: Hermenegildo MV, Cabeza D, editors. Practical Aspects of Declarative Languages. Vol. 3350. Berlin, Heidelberg: Springer; 2005. pp. 22-36. DOI: 10.1007/978-3-540-30557-6_4

[68] El-Anwar O, Chen L. Maximizing the computational efficiency of temporary housing decision support following disasters. Journal of Computing in Civil Engineering. 2014; **28**(1):113-123. DOI: 10.1061/(ASCE)CP.1943-5487.0000244

[69] Bonami P, Kilinç M, Linderoth J. Algorithms and software for convex mixed integer nonlinear programs. In: Lee J, Leyffer S, editors. Mixed Integer Nonlinear Programming. New York, NY: Springer; 2012. pp. 1-39. DOI: 10.1007/978-1-4614-1927-3_1

[70] Šilih S, Žula T, Kravanja S. The MINLP optimization in civil engineering. In: Proceedings of the 9th WSEAS International Conference on Mathematical and Computational Methods in Science and Engineering; 5-7 November 2007; Trinidad and Tobago. 2007

[71] Bussieck MR, Pruessner A. Mixed-integer nonlinear programming. SIAG/OPT Views and News. 2003;**14**(1):19-22

[72] Lu Y, Wang S, Sun Y, Yan C. Optimal scheduling of buildings with energy generation and thermal energy storage under dynamic electricity pricing using mixed-integer nonlinear programming. Applied Energy. 2015; **147**:49-58. DOI: 10.1016/j.apenergy.2015.02.060

[73] Shabani N, Sowlati T. A mixed integer non-linear programming model for tactical value chain optimization of a wood biomass power plant. Applied Energy. 2013;**104**:353-361. DOI: 10.1016/j.apenergy.2012.11.01

[74] Fan Y, Xia X. A multi-objective optimization model for building envelope retrofit planning. Energy Procedia. 2015;**75**:1299-1304. DOI: 10.1016/j.egypro.2015.07.193

[75] Karmellos M, Kiprakis A, Mavrotas G. A multi-objective approach for optimal prioritization of energy efficiency measures in buildings: Model, software and case studies. Applied Energy. 2015;**139**:131-150. DOI: 10.1016/j.apenergy.2014.11.023

[76] Mungle S, Benyoucef L, Son YJ, Tiwari MK. A fuzzy clustering-based genetic algorithm approach for time-cost-quality trade-off problems: A case study of highway construction project. Engineering Applications of Artificial Intelligence. 2013;**26**(8):1953-1966. DOI: 10.1016/j.engappai.2013.05.006

[77] Monghasemi S, Nikoo MR, Khaksar Fasaee MA, Adamowski J. A novel multi criteria decision making model for optimizing time-cost-quality trade-off problems in construction projects. Expert Systems With Applications. 2015;**42**(6):3089-3104. DOI: 10.1016/j.eswa.2014.11.032

[78] Brownlee AEI, Wright JA. Constrained, mixed-integer and multi-objective optimisation of building designs by NSGA-II with fitness approximation. Applied Soft Computing. 2015;**33**:114-126. DOI: 10.1016/j.asoc.2015.04.010

[79] Xu J, Tu Y, Zeng Z. A nonlinear multiobjective bilevel model for minimum cost network flow problem in a large-scale construction project. Mathematical Problems in Engineering. 2012. 40 pages. DOI: 10.1155/2012/463976. Article ID: 463976

[80] Xu J, Tu Y, Lei X. Applying multiobjective bilevel optimization under fuzzy random environment to traffic assignment problem: Case study of a large-scale construction project. Journal of Infrastructure Systems. 2014; **20**(3):05014003. DOI: 10.1061/(ASCE)IS.1943-555X.0000147

[81] Zhang L, Du J, Zhang S. Solution to the time-cost-quality trade-off problem in construction projects based on immune genetic particle swarm optimization. Journal of Management in Engineering. 2014;**30**(2):163-172. DOI: 10.1061/(ASCE)ME.1943-5479.0000189

[82] Yang IT, Lin YC, Lee HY. Use of support vector regression to improve computational efficiency of stochastic time-cost trade-off. Journal of Construction Engineering and Management. 2014;**140**(1):04013036. DOI: 10.1061/(ASCE)CO.1943-7862.0000784

[83] Futrell BJ, Ozelkan EC, Brentrup D. Bi-objective optimization of building enclosure design for thermal and lighting performance. Building and Environment. 2015;**2**:591-602. DOI: 10.1016/j.buildenv.2015.03.039

[84] Yahya M, Saka MP. Construction site layout planning using multi-objective artificial bee colony algorithm with levy flights. Automation in Construction. 2014;**38**(3):14-29. DOI: 10.1016/j.autcon.2013.11.001

[85] Ning X, Lam KC, Lam MCK. Dynamic construction site layout

planning using max-min ant system. Automation in Construction. 2010; **19**(1):55-65. DOI: 10.1016/j. autcon.2009.09.002

[86] Tran DH, Cheng MY, Cao MT. Hybrid multiple objective artificial bee colony with differential evolution for the time-cost-quality tradeoff problem. Knowledge-Based Systems. 2015;**74**: 176-186. DOI: 10.1016/j. knosys.2014.11.018

[87] Marzouk M, Abdelhamid M, Elsheikh M. Selecting sustainable building materials using system dynamics and ant colony optimization. Journal of Environmental Engineering and Landscape Management. 2013; **21**(4):237-247. DOI: 10.3846/ 16486897.2013.788506

[88] Wang B, Xia X. Optimal maintenance planning for building energy efficiency retrofitting from optimization and control system perspectives. Energy and Buildings. 2015;**96**:299-308. DOI: 10.1016/j. enbuild.2015.03.032

Chapter 5

On the Practical Consideration of Evaluators' Credibility in Evaluating Relative Importance of Criteria for Some Real-Life Multicriteria Problems: An Overview

Maznah Mat Kasim

Abstract

A multicriteria (MC) problem usually consists of a set of predetermined alternatives or subjects to be analyzed, which is prescribed under a finite number of criteria. MC problems are found in various applications to solve various area problems. There are three goals in solving the problems: ranking, sorting or grouping the alternatives according to their overall scores. Most of MC methods require the criteria weights to be combined mathematically with the quality of the criteria in finding the overall score of each alternative. This chapter provides an overview on the practical consideration of evaluators' credibility or superiority in calculating the criteria weights and overall scores of the alternatives. In order to show how the degree of credibility of evaluators can be practically considered in solving a real problem, a numerical example of evaluation of students' academic performance is available in the Appendix at the end of the chapter. The degree of credibility of teachers who participated in weighting the academic subjects was determined objectively, and the rank-based criteria weighting methods were used in the example. Inclusion of the degree of credibility of evaluators who participated in solving multicriteria problems would make the results more realistic and accurate.

Keywords: multicriteria problem, credibility, weights, subjective, aggregation

1. Introduction

Multicriteria decision-making (MCDM) is now considered as one discipline of knowledge, which has been expanding very fast in its own domain. Basically, it is about how to make decision when the undertaken issue is surrounded with a multiple number of criteria. The MC problem consists of two main components, alternatives and criteria. In real-life situations, the alternatives are options, organizations, people, or units to be analyzed which are prescribed under a set of finite criteria or attributes. If the number of alternatives is finite and known, the task is to select the best or the optimal alternative, to rank the alternatives according to their overall quality or performance, or to sort or group the alternatives based on certain measurements or values. In this case, the MC problem is usually called as a multiattribute decision-making (MADM) problem, and the alternatives are prescribed under a finite number of criteria or attributes [1]. The MADM methods are utilized to handle discrete MCDM problems [2]. This chapter focuses on MADM problems or more generally MCDM problems, where this type of problem has a finite number of predetermined alternatives, which is described by several criteria or attributes. MCDM problems can be found in various sectors.

1.1 Examples of multicriteria decision-making problems

Selection problems are really of an MCDM type, a simple problem that we are facing almost every day, for example, when we want to select a dress or a shirt to wear. A decision to choose which dress or shirt is based on certain attributes or factors, such as for what function (office, leisure, and business), color preference, and style or fashion. Here, the types of dress/cloth are the alternatives, while all factors that become the basis of evaluation are the attributes. Another example is when we want to choose the best location to set up projects such as housing, industrial, agricultural activities, recreation center, hoteling, and so on. Many factors or criteria that may be conflicting with each other should be considered by the decision-makers. Selecting the best candidate for various positions that can be conducted in many settings such as face-to-face interviews or online test is also an MCDM problem since the selection will be based on certain requirements. Selecting employees in different organizations with different scope of jobs with different requirements imposed by the related organization can also be categorized as an MCDM problem.

Another example is about selection of the best supplier of a manufacturing firm [3, 4], selection of the best personal computer [5], and selection of a suitable e-learning system [6] to be implemented in an educational institution. These studies focused on selecting the best alternative from a finite number of alternatives that were prescribed under a few evaluation criteria. These studies have the same main issue that is the relative importance or the weights of the evaluation criteria toward the overall performance of the alternatives under study. The studies provide ways to find weights subjectively and how to aggregate the weights when a group of decision-makers were involved in judging the importance of the criteria.

In addition, conducting an evaluation of a program, for example, is usually done after identifying the aspects of the program to evaluate. We may have many programs to be evaluated under several aspects of evaluations with the involvement of one evaluator or a group of evaluators. In a different situation, it may be only one program to be evaluated under several aspects and may be evaluated by one or many evaluators. Besides, many other evaluation situations are usually performed with the presence of many criteria such as evaluation of students, evaluation of employees' performance, evaluation of learning approaches [7], and evaluation of students' performance [8]. In relation to the study about the evaluation of students' academic performance in primary schools, five academic subjects were assumed to have different contribution toward the overall performance of the students. A few experienced teachers were asked to evaluate the degree of importance of the subjects. The resulting weights of the academic subjects were incorporated in finding the overall academic performance of the students in year six in one selected primary school in the northern part of Malaysia. For the purpose of illustrating the practical consideration of the credibility of the evaluators, the problem of evaluation of students' academic performance is extended by considering the credibility of the teachers who participated in weighting the academic subjects. The detailed discussion is available in the Appendix at the end of the chapter.

1.2 Credibility of the evaluators

Referring to those examples of MCDM scenarios, decision-maker(s) or evaluator (s) are involved in many stages of the evaluation process in searching for the optimal solution. As all MCDM problems have two main components, the alternatives and the criteria or attributes, the decision-maker(s) or the evaluator(s) would involve in at least two situations: deciding the quality of each alternative based on each of the criteria and also finding the relative importance of the criteria toward the overall performance of the alternatives. As what is usually arose in solving MCDM problems, criteria are contributing at different level of importance and should become a concern to the decision-maker(s) or evaluator(s). The criteria or attributes of the units to be analyzed should not be assumed to have same contribution toward the overall quality of the alternatives.

Besides having a challenge in finding the suitable evaluator(s) or decision-maker(s), since they might come with different background and experience, they also come with different levels of superiority or credibility that should be taken into consideration. This issue should be thought seriously because the results may be misleading if those who are involved in doing the evaluation or judgment do not have enough experience or less credible to give judgment regarding the MCDM problem under study. Moreover, the results may differ among the evaluators if the evaluators are at different levels of superiority [9]. Therefore, the credibility of expert(s) or evaluator(s) or decision-maker(s) who are involved in assessing quality of the alternatives or relative importance of attributes should be taken into consideration.

Webster's New World College Dictionary defines credibility as the quality of being trustworthy or believable. Credibility is also interpreted by good reputation, reputation, honor, and the presence of someone who stands out in the professional community [10]. Meanwhile, professionalism refers to competence or skill expected of a professional. In other words, a professional is someone who is skilled, reliable, and entirely responsible for carrying out their duties and profession [11]. This definition of professionalism has a resemblance to the term of credibility so that the two are like two sides of a coin that cannot be separated. For the purpose of assessment or evaluation, professionalism and credibility are the competencies of assessors in carrying out their functions and roles well, full of commitment, trustworthiness, and accountability.

It is normal that the assessors have different levels of credibility, and their credibility should be considered together with their assessments or evaluations. This chapter provides an overview of the current work on how the credibility of the decision-maker(s) or evaluator(s) could be considered especially on evaluating the importance of the criteria or attributes of any MCDM under investigation, how to quantify the credibility of those people, and how that quantitative values could be incorporated in finding the overall score of the alternatives. This issue falls under the concept of group decision-making and extends it with the consideration of the degree of superiority or credibility of the decision-maker(s) or evaluator(s). By deliberation of different relative importance of the attributes plus the different level of credibility or superiority of those who are involved in finding the optimal solution of the MCDM problem, the solution of the problem would be more realistic, accurate, and representative of the true setting of the problem.

In achieving the objective of the writing, the chapter is organized as follows. The next section describes the basic notations for this chapter. Section 3 discusses the concept of weights and the related methods, particularly the rank-based weighting method. Section 4 discusses on the aggregation of criteria weights and the values of criteria. Section 5 explains how to aggregate the credibility of the evaluators who are involved in weighting or finding weights or relative importance of the criteria. Furthermore, Section 5 also illustrates two approaches to aggregate the degree of credibility of evaluators in finding the relative importance in order to find the overall performance of the alternatives and their rankings. Section 6 suggests a few ways to quantify the credibility of the evaluators. The conclusion of the chapter is in Section 7, which is followed by a list of all references of the chapter. A numerical example is provided in the Appendix at the end of the chapter.

2. Basic notation

Let $A = \{A_1, \dots, A_n\}$ be a set of n alternatives that are prescribed under m criteria, $C = \{C_1, \dots, C_m\}$, and x_{ij} be a value of alternative i, under criterion j, where $i = 1, \dots, n$ and $j = 1, \dots, m$. Let $w = \{w_1, \dots, w_m\}$ be the weight of the criteria with conditions that $0 \leq w_j \leq 1$ and $\sum_{j=1}^{m} w_j = 1$. This information can be illustrated as a decision matrix as shown in **Figure** 1.

In relation to the numerical example in the Appendix, the students are the alternatives, while the academic subjects are the criteria. So, $A = \{A_1, \dots, A_{10}\}$ represents a set of 10 students that are prescribed under five academic subjects, $C = \{C_1, \dots, C_5\}$, and x_{ij} is the score of student i, under academic subject j, where $i = 1, \dots, 10$ and $j = 1, \dots, 5$. The weights of the criteria, $w = \{w_1, \dots, w_5\}$, obviously refer to the relative importance of the academic subjects toward the composite or final score of each student.

Weights of Criteria	w_1	w_2	...	w_m
Alternatives/ Criteria	C_1	C_2	...	C_m
A_1	x_{11}	x_{12}	...	x_{1m}
A_2	x_{21}	x_{22}	...	x_{2m}
$\vdots$	$\vdots$	$\vdots$	...	$\vdots$
A_n	x_{n1}	x_{n2}	...	x_{nm}

Figure 1.
A multiattribute problem as a decision matrix.

3. Weights of criteria

In finding the relative importance of the criteria or simply the weights of the criteria, $w = \{w_1, \dots, w_m\}$, there are many methods available in literature which are classified into two main approaches, objective methods and subjective methods

[12]. The objective methods are data-driven methods where quality values of the criteria should be available prior to the evaluation of criteria's relative importance. Based on the criteria's values, proxy measures such as standard deviation, correlation, variance, range, coefficient of variation, and entropy [13–17] would represent the criteria weights to be calculated. In relation to the concept of entropy, it was introduced in the communication theory, usually refers to uncertainty. The measure of entropy is often used to quantify the information or message. However, the entropy measure has become the proxy measures of criterion weights in MCDM domain. In other words, these objective methods produce weights of criteria based on the intrinsic information of the criteria. These methods do not require evaluators to do the criteria weighting. No further discussion is included in this chapter because objective weights are not the focus of the chapter.

3.1 Rank-based weighting methods

This subsection focuses on the discussion of rank-based weighting methods [18, 19] as these methods are used in this chapter in the illustration of practical consideration of evaluators' credibility in evaluating relative importance of criteria for some real-life multicriteria problems. These methods are very easy to use but have good impact [20]. Three popular rank-based methods are rank-sum (RS), rank reciprocal (RR), and rank order centroid (ROC). The mathematical representations of the three methods are as follows.

Suppose r_j be a ranking of criterion j given by an evaluator where r_j is an integer number with possible values from 1 to m. The smaller value of r_j means that the ranking of that criterion is higher and more important than the other criteria. The value of $r = \{r_1, \dots, r_m\}$ can be transformed into $w = \{w_1, \dots, w_m\}$ by using any of the following formula for RS, RR, and ROC, respectively. It should be noted that the sum of weights of the criteria is usually equal to one:

$$w_{j(rs)} = \frac{2(m + 1 - r_j)}{m(m - 1)} \tag{1}$$

$$w_{j(rr)} = \frac{1/r_j}{\sum_{j=1}^{m} 1/r_j} \tag{2}$$

$$w_{j(roc)} = \frac{1}{m} \sum_{k=1}^{m} \frac{1}{r_k} \times I(r_k > r_j) \tag{3}$$

where $I(r_k > r_j) = \begin{cases} 1 \text{ if } r_k \geq r_j \\ 0 \text{ if} r_k < r_j \end{cases}$.

Referring to the numerical example in the Appendix, there are five criteria representing five academic subjects; r_j is a ranking of academic subject$j$ where r_j is an integer number with possible values from 1 to 5, while the value of $r = \{r_1, \dots, r_5\}$ represents ranks of academic subjects 1 to 5 that can be transformed into weights of academic subjects 1 to 5, $w = \{w_1, \dots, w_5\}$, respectively.

Many studies were conducted to study the performance of these rank-based methods as criteria weighting methods. For example, a simulation experiment was conducted on investigating the performance of the three rank-based weighting methods (RS, RR, RS) and equal weights (EW) where the data was generated on a random basis [16]. Three performance measures of the methods were "hit rate," "average value loss," and "average proportion of maximum value range achieved." The results show that the ROC was found to be the best technique in most cases an in every measure. Another study on these three rank-based weighting techniques and EW concludes that the rank-based methods have higher correlations with the so-called true weights than EW [21].

A study is also done where EW, RS, and ROC methods were compared to direct rating and ratio weight methods [22]. Basically, the direct rating method is a simple type of weighting approach in which the decision-maker or the evaluator must rate all the criteria according to their importance. The evaluator can directly quantify their preference of the criteria. The rating does not constrain the decision-maker's responses since it is possible for the evaluator to alter the importance of one criterion without adjusting the weight of another [23]. The comparison was conducted under a condition that the evaluators' judgments of the criteria weights are not certain and subject to random errors. The results show that the direct rating tends to give better quality of decision results when the uncertainty is set as small, while ROC provides comparable results to the ratio weights when a large degree of error is placed. Please note that the ratio weight method requires the evaluators firstly rank the related criteria based on their importance. The evaluators should allocate certain value such as 10 for the least important attribute, and the rest of attributes are judged as multiples of 10. The weight of a criterion is obtained by dividing the criterion's weight with the sum of all attributes' weights.

The superiority of ROC over other rank-based methods is also subsequently confirmed in different simulation conditions [24]. An investigation on RS, RR, and ROC weighting methods was also carried out by changing the number of criteria from two to seven [25]. It is found that ROC gives the largest gap between the weights of the most important criterion and the least. RS provides the flattest weight function in the linear form. For RR, the weight of the most important one descends most aggressively to that of the second highest weight value, and then, the function continues to move flatter. In relation to rank-based weighting methods, another rank-based method was proposed [26]. This new rank-based method is called as generalized sum of ranks (GRS). Further investigation was carried out where the performance of GRS was compared to RS, RR, and ROC using a simulation experiment. The result of the investigation shows that GRS has a similar performance to ROC.

Based on the previous discussion, it can be concluded that the three rank-based weighting methods, RS, RR, and ROC, are having good features especially the ROC method. Therefore, these rank-based methods are used in the current study to illustrate how to include the degree of credibility of the evaluators who are involved in ranking the importance of the criteria. Furthermore, converting the ranks into weight values is not difficult, and the related formula is given as in Equations (1), (2), and (3).

3.2 Other subjective weighting methods

Other subjective weighting methods are analytic hierarchy process (AHP) [4, 27, 28], swing methods [29, 30], graphical weighting (GW) method [31], and Delphi method [32]. The AHP technique was introduced in 1980 [33]. It is a very popular MC approach, and it is done by conducting pairwise comparison of the importance of each pair of criteria. A prioritization procedure is implemented to draw a corresponding priority vector, where this priority vector represents the criteria weights. Thus, if the judgments are consistent, all prioritization procedures would give the same results. At the same time, if the judgments are inconsistent, prioritization procedures will provide different priority vectors [34]. Nevertheless, AHP is widely criticized for being such a tedious process, especially when there are a significant number of criteria or alternatives.

For the swing method, the evaluator must identify an alternative with the worst consequences on all attribute. The evaluator(s) can change one of the criteria from the worst consequence to the best. Then, the evaluator(s) is asked to choose the criteria that he/she would most prefer to modify from its worst to its best level, the criterion with the most chosen swing is the most important, and 100 points is allocated to the most important criterion.

The GW method begins with a horizontal line that is marked with a series of number, such as (9-7-5-3-1-3-5-7-9). The evaluator is expected to place a mark that represents the relative importance of a criterion on the horizontal line with the basis that a criterion is either more, equally, or less important than another criterion by a factor of 1–9. Then, a decision matrix is built as a pairwise comparison matrix. A quantitative weight for a criterion can be calculated by taking the sum of each row, and then the scores are normalized to obtain an overall weight vector. The GW method enables the evaluators to express preferences in a purely visual way. However, GW is sometimes criticized, since it allows evaluator(s) to assign weights in a more relaxed manner.

A Delphi subjective weighting method [35] requires one focus group of evaluators to evaluate the relative importance of the criteria. Each evaluator remains nameless to each other that can reduce the risk of personal effects or individual bias. The evaluation is conducted in more than one round until the group ends with a consensus of opinions on the relative importance of the criteria under study. The main advantage of this method is that the method avoids confrontation of the experts [36]. However, to pool up such a focus group is quite costly and timely.

4. Aggregation of criteria weights and values of criteria

Finding the final score of each alternative is very important since the final scores of the alternatives are required to rank the alternatives. Basically, those alternatives with higher scores should be positioned at higher rankings and vice versa. In order to find the overall or composite or final values of each alternative, the criteria weights should be aggregated with each alternative's values of the corresponding criteria. There are many aggregation methods available in literature. The section focuses on simple additive weighted average (SAW) method as the chapter uses SAW in the numerical example (in the Appendix at the end of the chapter). Furthermore, SAW method is a very well-established method and very easy to use [16].

4.1 Simple additive weighted average (SAW) method

The mathematical equation for SAW is given as follows:

$$Score\ A_i = \sum_{j=1}^{m} w_j x_{ij} \tag{4}$$

$Score\ A_i$ is the overall score of alternative i. Based on $Score\ A_i$, where $i = 1, \dots, n$, the n alternatives could be ranked, selected, or sorted with the condition that the alternatives with the higher overall scores should be ranked at higher positions. Referring to the numerical example in the Appendix, $Score\ A_i$ represents the overall score of student i, where $i = 1, \dots, 10$.

SAW is an old method, and MacCrimmon is one of the first researchers that summarized this method in 1968 [37]. As a well-established method, it is used widely [38] in solving MC problems, particularly for the evaluation of alternatives. Basically, this method is the same as the simple arithmetic average method, but instead of having the same weight values for the criteria, SAW method uses mostly distinct weights values of the criteria. As given in Eq. (4), the overall performance of each alternative is obtained by multiplying the rating of each alternative on each criterion by the weight assigned to the criterion and then summing these products over all criteria [15]. The best alternative is the one that obtained the highest score and will be selected or ranked at the first position. Many recent studies used the SAW method, for example, in [39–41], and a review on its applications is also available [42].

Besides SAW or also known as weighted sum method (WSM), there is another average technique, called weighted product model (WPM) or simple geometric weighted (SGW) or simple geometric average method. In WPM, the overall performance of each alternative is determined by raising the rating of the alternative to the power of the criterion weight and then multiplying these products over all criteria [15]. However, WPM is a little bit complex as compared to SAW since WPM involves power and multiplications.

4.2 Other aggregation methods

AHP [14], technique for order preference by similarity to ideal solution (TOPSIS), and *VlseKriterijumska Optimizacija Kompromisno Resenje* (VIKOR) [43] are also popular aggregation methods in solving MC problems. As previously mentioned in Section 3.2, AHP is built under the concept of pairwise comparison either in finding the criteria weights or criteria values of the alternatives. The aggregation of criteria weights and the criteria values obtained by AHP is sometimes done by using the SAW or SGW methods.

AHP and TOPSIS are two different aggregation methods. TOPSIS assigns the best alternative that relies on the concepts of compromise solution, where the best alternative is the one that has the shortest distance from the ideal solution and the farthest distance from the negative ideal solution [44]. In other words, alternatives are prioritized according to their distances from positive ideal solutions and negative ideal solutions, and the Euclidean distance approach is utilized to evaluate the relative closeness of the alternatives to the ideal solutions. There is a series of steps of TOPSIS, but this method starts with the weighted normalization of all performance values against each criterion. Some recent applications of the TOPSIS method are available [45–48].

VIKOR method [49] is quite similar to TOPSIS method, but there are some important differences, and one of the differences is about the normalization process. TOPSIS uses the vector linearization where the normalized value could be different for different evaluation unit of a certain criterion, while VIKOR uses linear normalization where the normalized value does not depend on the evaluation unit of a criterion. VIKOR has also been used in many real-world MCDM problems such as mobile banking services [50] digital music service platforms [51], military airport location selection [52], concrete bridge projects [53], risk evaluation of construction projects [54], maritime transportation [55], and energy management [56].

5. Inclusion of credibility of evaluators in solving multicriteria problems

This section discusses how credibility can be included practically in solving MC problems. Suppose the evaluators are requested to evaluate the relative importance

$$\begin{pmatrix} r_1^1, \dots, r_1^p \\ \vdots \\ r_m^1, \dots, r_m^p \end{pmatrix} \xrightarrow{r^l_{(rs)j}} \begin{pmatrix} u^1 w_1^1, \dots, u^1 w_1^p \\ \vdots \\ u^p w_m^p, \dots, u^p w_m^p \end{pmatrix} \xrightarrow{\sum_{l=1}^{p} \frac{u^l w_j^l}{p}} (w_1^{av}, \dots, w_m^{av}) \xrightarrow{Score\ A_i = \sum_{j=1}^{m} w_j^{av} x_{ij}} \begin{pmatrix} Score A_1 \\ \vdots \\ Score A_n \end{pmatrix} \rightarrow R$$

Figure 2.
Approach 1.

of the criteria based on rank-based weighting methods as explained in Section 3.1. Suppose there is a panel of p evaluators, and let r_j^l be rank of criterion j, evaluated by evaluator l, where $l = 1, \dots, p$. In order to include the credibility of the evaluators, let us introduce a new set of values that represents the different credibility of the evaluators. Let u^l be the degree of credibility of evaluator l, where $0 \leq u^l \leq 1$, and $\sum_{l=1}^{p} u^l = 1$. There are two approaches [57] where the degree of credibility of the evaluators could be attached in finding the overall scores. The first approach is in calculating the final weight of criteria as given in **Figure 1**, and the second approach is in computing the overall performance of the alternatives as given in **Figure 2**.

For the first approach as portrayed in **Figure 2**, the degree of credibility of the evaluators is attached to the resulted weights from the ranks of criteria by using any of the equations, Eq. (1), Eq. (2), or Eq. (3). So, here there are p sets of weights of the criteria, and the average of that p weights for each criterion is calculated by summing up all weights for that criteria and divide the sum with the total number of evaluators. So now, there is only one set of weights that can be aggregated with the values of alternatives for each corresponding criterion as given in Eq. (4). There is only one set of overall performance of all n alternatives.

For the second approach, the criteria weights obtained from each evaluator are kept, and then each set of weights is aggregated with the quality values of each alternative. So, here there are p sets of overall values of the alternatives. In order to get the final overall score of the alternatives, the average of the p scores for each alternative should be calculated. The ranking or sorting of the alternatives or selecting the best alternative is done based on the average of that p overall scores of each individual alternative. The following section provides some suggestion on how to quantify the credibility of the evaluators.

Referring to the numerical example in the Appendix, there were three evaluators involved in ranking the importance of the five academic subjects, and the number of students is 10. So, r_j^l is the rank of academic subject j, with $j = 1, \dots, 5$, evaluated by evaluator l, where $l = 1, \dots, 3$, and $n = 10$, while u^l represents the degree of credibility of evaluator l, where $0 \leq u^l \leq 1$, and $\sum_{l=1}^{3} u^l = 1$.

6. Quantification of credibility of evaluators

Credibility is synonym to professionalism, integrity, trustworthiness, authority, and believability. A study focuses on how to assess the credibility of expert witnesses [58]. A 41-item measure was constructed based on the ratings by a panel of judges, and a factor analysis yielded that credibility is a product of four factors: likeability, trustworthiness, believability, and intelligence. Another study concerns about the credibility of information in digital era [59]. Credibility is said to have two main components: trustworthiness and expertise. However, the authors conclude that the relation among youth, digital media, and credibility today is sufficiently complex to resist simple explanations, and their study represents a first step toward mapping that complexity and providing a basis for future work that seeks to find explanations.

It can be argued that the degree of credibility of evaluators or judges or decision-makers can be determined subjectively or objectively, where the former one can be done by using certain construct as proposed in [58] or can be determined based on certain objective or exact measures such as years of experience, salary scale, or amount of salary. The quantification of the degree of credibility opens a new potential area of research as there are very few researches done especially on finding the suitable objective proxy measures of the degree of credibility.

Finding the degree of credibility subjectively requires more time and much harder as it involves a construct or an instrument which would be used as a rating mechanism to obtain the degree of credibility. Meanwhile, finding the degree of credibility based on objective information is simpler and easier to do. As an illustration on how to quantify the credibility objectively, suppose there are three experts with their basic salaries in a simple ratio of 1:2:3. So, this ratio can be converted as 0.167:0.333:0.500, so that the sum of credibility of the evaluators is equal to 1. These values can be used to represent the degree of credibility of the evaluators or experts 1, 2, and 3, respectively. It should be noted that the sum of the degrees of credibility of the three evaluators is equal to one to make the future calculation simple while easier for interpretation of the values. Here, evaluator 3 is the most credible one since he/she has the highest salary among the three, and it is a usual practice that those who are higher in terms of expertise usually are paid higher. The same computation can be used for the years of experience or salary scale.

The numerical example in the Appendix extends the problem of evaluating students' academic performance which is discussed earlier in the Introduction. Here, the credibility of the teachers who were asked to assess the relative importance of the five subjects was considered. In order to incorporate the degree of credibility of the teachers, a new set of values is introduced to represent these different degrees of credibility. The example shows two ways of calculations on how the credibility values could be included in finding the overall scores of the alternatives. As expected, the overall scores and the overall ranking are different as compared to overall scores of not considering the different credibility of the teachers. The details and the step-by-step methodology are also included in the Appendix.

7. Conclusion

This chapter provides an overview on the practical consideration of evaluators' credibility in evaluating relative importance of criteria for some real-life multicriteria problems. Credibility of the evaluators who are involved in solving any multicriteria problem should be included in calculating the overall scores of the alternatives or the units of analysis. This chapter demonstrates how the credibility of evaluators who participated in finding the criteria weights can be combined with the criteria weights and the quality of the criteria of the alternatives. Rank-based criteria weighting methods are used as an illustration in a numerical example of evaluation of students' academic performance problem at the end of the chapter. However, other criteria subjective weighting methods are also possible to be used but with caution especially at the stage of aggregation of criteria weights and criteria values. It may exist only one approach to do the aggregation due to the underpinning concepts of the aggregation methods. The chapter uses simple additive weighted average method as the aggregation method since the method is very well established. The use of other aggregation techniques is also plausible. The chapter also suggests a few practical proxy measures of the credibility but is still very limited. More researches should be conducted to find ways of measuring the credibility of evaluators or experts either subjectively or objectively. Inclusion of the credibility of evaluators in solving multicriteria problems is realistic since the evaluators come from different backgrounds and levels of experience. Quantification of the evaluators' credibility subjectively or objectively opens a new insight in group decision-making field. Furthermore, the credibility of the evaluators should also be considered in other multicriteria problems in other areas, so that the results are more practical and accurate.

Appendix: A numerical example

Mr. Zachariah is a class teacher of 10 excellent students in one of the best primary schools of a country. The 10 students were already given the final marks of five main academic subjects by their respective teachers as in **Table 1**.
Mr. Zachariah must rank the students according to their performance because these students will be given awards and recognition on their graduation day.

Suppose three experienced teachers, Edward, Mary, and Foong, were asked to evaluate the relative importance of the five academic subjects with their degree of credibility as discussed in previous section, that is, the salary ratio of the three teachers is 0167: 0.333: 0.500. The rank-based technique is used to analyze the ranking of importance of the academic subjects given by these three teachers by using Eq. (1).

The results are given in **Table 2**. Column 2 displays the ranking of the criteria evaluated by teacher 1, and column 3 shows the corresponding criteria weights as analyzed by Eq. (1), while columns 4 and 5 and columns 6 and 7 show the respective results by teachers 2 and 3, respectively. The second last column of the table summarizes the criteria weights when the teachers are of same credibility. The values were computed as the simple arithmetic average of the corresponding criterion, while the last column has the final weights that were calculated as the simple arithmetic average as well but with consideration of the different degree of credibility according to Approach 1 as given in **Figure 2**. Please note that the both sets of final weights are already summed to one. So, the normalization process to guarantee the sum of weights is one and is not necessary.

	Native language	English language	Mathematics	Science	History
Student 1, A_1	0.25	0.34	0.12	0.36	0.45
A_2	0.33	0.54	0.22	0.44	0.76
A_3	0.43	0.65	0.57	0.42	0.91
A_4	0.55	0.32	0.37	0.67	0.53
A_5	0.27	0.66	0.57	0.82	0.61
A_6	0.67	0.56	0.46	0.46	0.31
A_7	0.58	0.87	0.39	0.27	0.43
A_8	0.32	0.76	0.41	0.37	0.51
A_9	0.91	0.36	0.47	0.45	0.45
A_{10}	0.12	0.33	0.81	0.75	0.32

Table 1.
Ten students assessed under five academic subjects.

	Teacher 1 (0.167)		Teacher 2 (0.333)		Teacher 3 (0.500)		Final weight same credibility (SC)	Final weight different credibility (DF)
	r^1	w^1	r^2	w^2	r^3	w^3		
Native language	1	0.333	2	0.267	2	0.267	0.289	0.278
English language	3	0.200	3	0.200	3	0.200	0.200	0.200
Mathematics	4	0.133	1	0.333	1	0.333	0.267	0.300
Science	5	0.067	5	0.067	5	0.067	0.067	0.067
History	2	0.267	4	0.133	4	0.133	0.178	0.156

Table 2.
Criteria weights of five academic subjects evaluated by three teachers with the same and different credibility by using rank-sum weighting technique.

Now, in order to find the overall performance of each student, for example, the overall performance of student 1 without consideration of credibility of teachers in evaluating the relative importance of the academic subjects, it is simply done by multiplying row 2 of **Table 1** with its corresponding criteria weights in the second last column of **Table 2** by using Eq. (4) as follows:

$$\begin{aligned}\text{Score } A_1 &= \sum_{j=1}^{5} w_j x_{1j} \\ &= (0.289)(0.25) + (0.2)(0.34) + (0.267)(0.12) + (0.067)(0.36) \\ &\quad + (0.178)(0.45) \\ &= 0.277\end{aligned}$$

The same process is performed to find the overall scores of student 1, if the credibility of the teachers in finding weights of the criteria is considered but the weights in last column of **Table 2** is used, instead.

$$\begin{aligned}\text{Score } A_1 &= \sum_{j=1}^{5} w_j x_{1j} \\ &= (0.278)(0.25) + (0.2)(0.34) + (0.3)(0.12) + (0.067)(0.36) \\ &\quad + (0.156)(0.45) \\ &= 0.244\end{aligned}$$

Table 3 gives the overall scores and the corresponding final rankings of all students based on average criteria weights with the same (SC) and different (DC) credibility of the teachers. The overall scores are all different, while the rankings are different especially for ranks 8 and 9 and 4 and 5.

Table 4 summarizes three individual overall score of the three different teachers without consideration of their credibility, while the second last column and the last column are the average overall scores of the three overall scores and its corresponding rankings, respectively.

Table 5 shows the three overall scores by consideration of the credibility of teachers in finding the academic subjects' weights, and the average overall scores of the three overall scores. The ranking of the students is based on the average overall scores in column 5 of the table. Here, Approach 2 as in **Figure 3** is used to find the final overall scores of the students.

To make the comparison easier, **Table 6** summarizes the overall scores and their corresponding rankings of the students with SC and DC of the teachers when calculating the academic subjects' weights based on Approach 2.

		A_1	A_2	A_3	A_4	A_5	A_6	A_7	A_8	A_9	A_{10}
SC	Score	0.277	0.427	0.597	0.461	0.526	0.514	0.540	0.470	0.571	0.424
	Rank	10	8	1	7	4	5	3	6	2	9
DC	Score	0.244	0.408	0.592	0.439	0.518	0.540	0.550	0.462	0.561	0.422
	Rank	10	9	1	7	5	4	3	6	2	8

Table 3.
Overall scores and ranking of students with average criteria weights evaluated by teachers of the same and different credibility based on Approach 1.

	Score A_i^1	Score A_i^2	Score A_i^3	Score A_i^{AV}	Ranking
A_1	0.311	0.259	0.259	0.276	10
A_2	0.479	0.400	0.400	0.426	8
A_3	0.620	0.584	0.584	0.596	1
A_4	0.483	0.449	0.449	0.460	7
A_5	0.515	0.530	0.530	0.525	4
A_6	0.510	0.516	0.516	0.514	5
A_7	0.552	0.534	0.534	0.540	3
A_8	0.474	0.467	0.467	0.469	6
A_9	0.588	0.561	0.561	0.570	2
A_{10}	0.349	0.461	0.461	0.424	9

Table 4.
Same credibility: four different sets of overall scores and final ranking of the 10 students based on average overall scores.

	u^1ScoreA_i^1	u^2Score A_i^2	u^3Score A_i^3	Score A_i^{AV}	Ranking
A_1	0.052	0.086	0.129	0.089	10
A_2	0.080	0.133	0.200	0.138	9
A_3	0.104	0.194	0.292	0.197	1
A_4	0.081	0.150	0.225	0.152	7
A_5	0.086	0.176	0.265	0.176	4
A_6	0.085	0.172	0.258	0.172	5
A_7	0.092	0.178	0.267	0.179	3
A_8	0.079	0.155	0.233	0.156	6
A_9	0.098	0.187	0.281	0.189	2
A_{10}	0.058	0.153	0.230	0.147	8

Table 5.
Different credibility: four different sets of overall scores and final ranking of the 10 students based on average overall scores.

As the two sets of the overall scores are different, all rankings based on both sets of the overall scores are the same except for ranks 8 and 9. There is not much different in the overall rankings since the MC problem that is considered here is

$$\begin{pmatrix} r_1^1, \dots, r_1^p \\ \vdots \\ r_m^1, \dots, r_m^p \end{pmatrix} \xrightarrow{r_{(rs)j}^l} \begin{pmatrix} w_1^1, \dots, w_1^p \\ \vdots \\ w_m^p, \dots, w_m^p \end{pmatrix} \xrightarrow{u^l Score\ A_i^l} \begin{pmatrix} u^1 Score A_1^1, \dots, u^1 Score A_n^1 \\ \vdots \\ u^p Score\ A_1^p, \dots, u^p Score\ A_n^p \end{pmatrix} \rightarrow \begin{pmatrix} Score A_1^{av} \\ \vdots \\ Score A_n^{av} \end{pmatrix} \rightarrow R$$

Figure 3.
Approach 2.

		A_1	A_2	A_3	A_4	A_5	A_6	A_7	A_8	A_9	A_{10}
SC	Score	0.276	0.426	0.596	0.460	0.525	0.514	0.540	0.469	0.570	0.424
	Rank	10	8	1	7	4	5	3	6	2	9
DC	Score	0.089	0.138	0.197	0.152	0.176	0.172	0.179	0.156	0.189	0.147
	Rank	10	9	1	7	4	5	3	6	2	8

Table 6.
Two different set of overall scores of the students by averaging overall performance of the students and their corresponding rankings based on Approach 2.

only a small scale problem with only 10 alternatives and 5 criteria. However, the two sets of overall values are totally different. There may be much more differences in terms of rankings if a bigger MC problem with more alternatives and more criteria is considered. The final ranking of the students obtained by consideration of the different credibility of the teachers should be selected as the practical and valid results.

Author details

Maznah Mat Kasim
School of Quantitative Sciences (SQS), UUM College of Arts and Sciences, Universiti Utara Malaysia, Sintok, Kedah, Malaysia

*Address all correspondence to: maznah@uum.edu.my

References

[1] Hwang CL, Paidy R, Yoon K, Masud AS. Mathematical programming with multiple objectives: A tutorial. Computer & Operations Research. 1980;7:5-31

[2] Rezaei J. Best-worst multi-criteria decision-making method. Omega. 2015; **53**:49-57

[3] Ahmad N, Kasim MM, Rajoo SSK. Supplier selection using a fuzzy multi-criteria method. International Journal of Industrial Management. 2016;**2**:61-71

[4] Ahmad N, Kasim MM, Ibrahim H. The integration of fuzzy analytic hierarchy process and VIKOR for supplier selection. International Journal of Supply Chain Management. 2017; **6**(4):289-293

[5] Kasim MM, Ibrahim H, Al-Bataineh MS. Multi-criteria decision-making methods for determining computer preference index. Journal of Information and Communication Technology. 2011;**10**:137-148

[6] Mohammed HJ, Kasim MM, Shaharanee INM. Selection of suitable e-learning approach using TOPSIS technique with best ranked criteria weights. In: Proceedings of the 13th IMT-GT International Conference on Mathematics, Statistics and Their Applications (ICMSA 2017); 4-7 December 2017; Malaysia. Sintok: AIP Conference Proceedings 1905. 2017. p. 040019-1-6

[7] Mohammed HJ, Kasim MM, Hamadi AK, Al-Dahneem E. Evaluating collaborative and cooperative learning using MCDM method. Advanced Science Letters. 2018;**24**(6):4084-4088

[8] Kasim MM, Abdullah SRG. Aggregating student academic achievement by simple weighted average method. Malaysian Journal of Learning and Instructions. 2013;**10**: 119-132

[9] Metzger MJ. Making sense of credibility on the Web: Models for evaluating online information and recommendations for future research. Journal of the American Society for Information Science and Technology. 2007;**58**:2078-2091

[10] Zulfiqar S, Bin Tahir S. Professionalism and credibility of assessors in enhancing educational quality. 2019;**2**:162-175. Available from: https://www.researchgate.net/publication/334599511_PROFESSIONALISM_AND_CREDIBILITY_OF_ASSESSORS_IN_ENHANCING_EDUCATIONAL_QUALITY [Accessed: 6 April 2020]

[11] Korten D, Alfonso F. Bureaucracy and the Poor: Closing the Gap. Singapore: McGraw-Hill; 1981

[12] Ma J, Fan Z, Huang L. A subjective and objective integrated approach to determine attribute weights. European Journal of Operational Research. 1999; **112**:397-404

[13] Ray AM. On the measurement of certain aspects of social development. Social Indicators Research. 1989;**21**:35-92

[14] Diakoulaki D, Koumoutos N. Cardinal ranking of alternatives actions: Extension of the PROMETHEE method. European Journal of Operational Research. 1991;**53**:337-347

[15] Hwang CL, Yoon K. Multiple Attribute Decision-Making: Methods and Applications. Berlin: Springer; 1981

[16] Triantaphyllou E. Multi Criteria Decision Making: A Comparative Study.

London: Kluwer Academic Publisher; 2000

[17] Zeleny M. Multiple Criteria Decision Making. New York: McGraw-Hill Book Company; 1982

[18] Barron FH, Barrett BE. Decision quality using ranked attribute weights. Management Science. 1996;**42**(11): 1515-1523

[19] Roberts R, Goodwin P. Weight approximation in multi-attribute decision models. Journal of Multi-Criteria Decision Analysis. 2002;**11**:291-303

[20] Desa NHM, Jemain AA, Kasim MM. Construction of a composite hospital admission index in Klang Valley, Malaysia by considering the aggregated weights of criteria. Sains Malaysiana. 2015;**44**(2):239-247

[21] Stillwell WG, Seaver DA, Edwards W. A comparison of weight approximation techniques in multi-attribute utility decision making. Organizational Behavior and Human Performance. 1981;**28**(1):62-77

[22] Jia J, Fischer GW, Dyer JS. Attribute weighting methods and decision quality in the presence of response error: A simulation study. Journal of Behavioral Decision Making. 1998;**11**(2):85-105

[23] Arbel A. Approximate articulation of preference and priority derivation. European Journal of Operational Research. 1989;**43**:317-326

[24] Ahn BS, Park KS. Comparing methods for multi-attribute decision making with ordinal weights. Computers & Operations Research. 2008;**35**(5):1660-1670

[25] Roszkowska W. Rank ordering criteria weighting methods—A comparative overview. Optimum Studia Ekonomiczne NR. 2013;**5**(65):14-33

[26] Wang J, Zionts S. Using ordinal data to estimate cardinal values. Journal of Multi-Criteria Decision Analysis. 2015; **22**(3-4):185-196

[27] Saaty TL. How to make decision: The analytical hierarchy process. Journal of Operational Research Society. 1990; **48**:9-26

[28] Mohammed HJ, Kasim MM, Shaharanee INM. Evaluating of flipped classroom learning using analytic hierarchy process technique. International Journal of Trend in Research and Development (IJTRD). 2017;**4**(2):443-446

[29] Von Winterfeldt D, Edwards W. Decision Analysis and Behavioral Research. Cambridge: University Press; 1986

[30] Edwards W. How to use multi attribute utility measurement for social decision making. IEEE Transaction on Systems Man and Cybernetics. 1977: 326-340

[31] Hajkowicz SA, McDonald GT, Smith PN. An evaluation of multiple objective decision support weighting techniques in natural resource management. Journal of Environmental Planning and Management. 2000;**43**(4): 505-518

[32] Chang YC, Hsu CH, Williams G, Pan ML. Low cost carriers' destination selection using a Delphi method panel. Tourism Management. 2008;**5**(29): 898-908

[33] Saaty TL. The Analytic Hierarchy Process. New York: McGraw-Hill; 1980

[34] Saaty RW. The analytic hierarchy process—What it is and how it is used. Mathematical Modelling. 1987;**9**(3-5): 161-176

[35] Still BG, May AD, Bristow AL. The assessment of transport impacts on land

use: Practical uses in strategic planning. Transport Policy. 1999;**6**(2):83-98

[36] Zardani NH, Ahmed K, Shirazi SM, Yusob ZB. Weighting Methods and Their Effects on Multi-Criteria Decision-Making Model Outcomes in Water Resources Management. Cham, Heidelberg, New York, Dordrecht, London: Springer; 2015

[37] MacCrimmon KR. Decision-Making Among Multiple-Attribute Alternatives: A Survey and Consolidated Approach. Santa Monica: Rand Corp Santa Monica; 1968

[38] Chen SJ, Hwang CL. Fuzzy multiple attribute decision making methods. In: Fuzzy Multiple Attribute Decision Making. Berlin, Heidelberg: Springer; 1992

[39] Wang LE, Liu HC, Quan MY. Evaluating the risk of failure modes with a hybrid MCDM model under interval-valued intuitionistic fuzzy environments. Computers & Industrial Engineering. 2016;**102**:175-185

[40] Kalibatas D, Kovaiti V. Selecting the most effective alternative of waterproofing membranes for multifunctional inverted flat roofs. Journal of Civil Engineering and Management. 2017;**23**(5):650-660

[41] Muddineni VP, Sandepud SR, Bonala AK. Improved weighting factor selection for predictive torque control of induction motor drive based on a simple additive weighting method. Electric Power Components and Systems. 2017; **45**(13):1450-1462

[42] Abdullah L, Adawiyah CR. Simple additive weighting methods of multi criteria decision making and applications: A decade review. International Journal of Information Processing and Management. 2014;**5**(1): 39-49

[43] Opricovic S. Multicriteria Optimization of Civil Engineering Systems. Belgrade: Faculty of Civil Engineering; 1998

[44] Tzeng GH, Huang JJ. Multiple Attribute Decision-making: Methods and Applications. New York: Chapman and Hall/CRC; 2011

[45] Mao N, Song M, Pan D, Deng S. Comparative studies on using RSM and TOPSIS methods to optimize residential air conditioning systems. Energy. 2018; **144**:98-109

[46] Chen W, Shen Y, Wang Y. Evaluation of economic transformation and upgrading of resource-based cities in Shaanxi province based on an improved TOPSIS method. Sustainable Cities and Society. 2018;**37**:232-240

[47] Shen F, Ma X, Li Z, Xu Z, Cai D. An extended intuitionistic fuzzy TOPSIS method based on a new distance measure with an application to credit risk evaluation. Information Sciences. 2018;**428**:105-119

[48] Polat G, Eray E, Bingo BN. An integrated fuzzy MCGDM approach for supplier selection problem. Journal of Civil Engineering and Management. 2017;**23**(7):926-942

[49] Opricovic S, Tzeng GH. Compromise solution by MCDM methods: A comparative analysis of VIKOR and TOPSIS. European Journal of Operational Research. 2004;**156**(2): 445-455

[50] Lu MT, Tzeng GH, Cheng H, Hsu CC. Exploring mobile banking services for user behavior in intention adoption: Using new hybrid MADM model. Service Business. 2014;**9**(3): 541-563

[51] Lin CL, Shih YH, Tzeng GH, Yu HC. A service selection model for digital music platforms using a hybrid MCDM

approach. Applied Soft Computing. 2016;**48**:385-403

[52] Sennaroglu B, Varlik Celebi G. A military airport location selection by AHP integrated PROMETHEE and VIKOR methods. Transportation Research Part D: Transport and Environment. 2018;**59**:160-173

[53] Gao Z, Liang RY, Xuan T. VIKOR method for ranking concrete bridge repair projects with target-based criteria. Results in Engineering. 2019;**3**: 100018

[54] Wang L, Zhang HY, Wang JQ, Li L. Picture fuzzy normalized projection based VIKOR method for the risk evaluation of construction project. Applied Soft Computing. 2018;**64**: 216-226

[55] Soner O, Celik E, Akyuz E. Application of AHP and VIKOR methods under interval type 2 fuzzy environment in maritime transportation. Ocean Engineering. 2017;**129**:107-116

[56] Sakthivel G, Sivakumar R, Saravanan N, Ikua BW. A decision support system to evaluate the optimum fuel blend in an IC engine to enhance the energy efficiency and energy management. Energy. 2017;**140**: 566-583

[57] Kasim MK, Jemain AA. Involvement of panel of evaluators with different credibility in aggregating subjective rank-based values. Sains Malaysiana. 2013;**42**(5):667-672

[58] Flanagin AJ, Metzger MJ. Digital media and youth: Unparalleled opportunity and unprecedented responsibility. In: Metzger M, Flanagin A, editors. Digital Media, Youth, and Credibility. Cambridge: The MIT Press; 2008

[59] Brodsky SL, Griffin MP, Cramer RJ. The witness credibility scale: An outcome measure for expert witness research. Behavioral Sciences & the Law. 2010;**28**(6):892-907